CHEMICAL COMMUNICATION

Language of Chemistry

Dr. Shailesh K Jain

Professor

Department of Applied Sciences and Humanities

KCC Institute of Technology and Management

Knowledge Park-III, Greater Noida

Every possible effort has been made to ensure that the information contained in this book is accurate at the time of going to press, and the publisher and author cannot accept responsibility for any errors or omissions, however caused. No responsibility for loss or damage occasioned to any person acting, or refraining from action, as a result of the material in this publication can be accepted by the editor, the publisher or the author.

Every effort has been made to trace the owners of copyright material used in this book. The author and the publisher will be grateful for any omission brought to their notice for acknowledgement in the future editions of the book.

Copyright © Author, 2018

All rights reserved. No part of this book may be reproduced, stored in a retrieval system, or transmitted in any form or by any means, electronic, mechanical, photocopying, recorded or otherwise, without the written permission of the publisher.

MV Learning
A Viva Books imprint

3, Henrietta Street
London WC2E 8LU
UK

4737/23, Ansari Road,
Daryaganj, New Delhi 110 002
India

ISBN: 978-93-87486-36-2

Printed and bound in India.

Preface

Communication is very important in every walk of life. All animals and human beings communicate with each other in their own language.

The systematic mode of expressing ones' view to others and understanding those of others is called language. Like other languages, the chemical language has also evolved and grown with a developed vocabulary and language for scientific communication.

The basic alphabet in chemical language is called the **identifier**. The identifier for an element is its symbol, for a compound it is the formula and for a chemical reaction it is the chemical equation.

In conventional languages, letters give rise to words, words to sentences and sentences to paragraphs. Similarly, symbols for elements combine to give formulae, the formulae when combined give chemical equations and the chemical equations describe the chemical reactions. Thus you see that the chemical language is not much different from the conventional language we speak or write.

WHY THIS BOOK?

The idea of preparing such a supplementary study material popped up during interactions with teachers at seminars and workshops across the country. During such interactions, it was realised that the students face difficulties in writing formulae, chemical equations and more so in naming compounds and balancing chemical equations.

FOR WHOM IS THIS BOOK?

The book, **Chemical Communication**, has been prepared for

- All the students who need or wish to study fundamentals of chemistry.
- Those who did not have the opportunity to study chemistry during their formative schooling.
- Those who have studied chemistry in their junior classes and wish to reinforce their basic knowledge and understanding of this all important language.
- Teachers and parents with limited background of chemistry, but who are called on to teach the subject.

SALIENT FEATURES OF THE BOOK

The basic objective of preparing this study material is to familiarise the beginners to perform the following:

- Write the symbols of various elements, formulae of elements and compounds and their names and vice-versa.
- Describe the elements in terms of symbols, compounds in terms of formulae and chemical reactions in terms of chemical equations or ionic equations.

- Write chemical equations from the word equations and vice-versa.
- Balance chemical equations (both molecular and ionic) by various methods.
- Name Acids, Bases and Salts from their formulae and vice-versa.
- Name simple organic compounds on the basis of IUPAC recommendations.
- Test their understanding by working out a large number of questions included in the text.

I hope my fellow teachers and the students for whom this book has been written will find it both interesting and useful.

I express my sincere thanks to all the students and teachers for their constructive feedback over the years and continued support to make this book more user-friendly study material.

Dr. Shailesh K Jain
shailejain@gmail.com

Contents

1

Matter and Its Classification

MATTER – Matter is the material, which the universe is composed of. Matter is something that occupies space, possesses mass, and can be felt by one or more of our senses.

Water, Air, Wood, Sugar, Cotton, Book, etc. consist of matter.

A definite variety of matter, all samples of which have the same properties, is commonly known as a **substance.** For example, Sugar is a substance because even the smallest particle of sugar is sweet.

MATTER CLASSIFIED ON THE BASIS OF ITS PHYSICAL STATES

SOLID – The matter that occupies a definite space and has definite shape is called solid.

LIQUID – The matter that occupies a definite space but has no definite shape is called liquid.

GAS – The matter that neither occupies a definite space nor has a definite shape is called gas.

Water can exist as ice (solid state), water (liquid state) and steam (gaseous state).

MATTER					
Solid		Liquid		Gas	
Ice	Sand	Water	Milk	Air	Hydrogen
Sugar	Diamond	Oil	Mercury	Oxygen	Nitrogen
Gold	Silver	Kerosene	Petrol	Carbon dioxide	Helium
Iron	Wood	Alcohol	Bromine	Chlorine	Argon

CONSTITUENTS OF MATTER

ATOM – An atom is the smallest particle of an element that takes part in chemical reactions.

MOLECULE – The smallest unit of a substance which is capable of independent existence and shows all the properties of that substance is called a molecule.

An atom maintains its chemical identity throughout all chemical and physical changes.

- A molecule contains two or more atoms of the same or different types.
- Molecule of the noble gases, e.g., Helium, Neon, Argon, Krypton and Xenon, contains only one atom.

MATTER CLASSIFIED ON THE BASIS OF ITS CHEMICAL CONSTITUTION

On the basis of its chemical constitution, matter is classified into two kinds:

- Elements
- Compounds

An element is composed of atoms of the same kind.

ELEMENTS – An **element** is the simplest form of matter, which cannot be split into simpler substances by any chemical or physical method.

ELEMENTS					
Metals		**Nonmetals**		**Metalloids**	
Gold	Iron	Hydrogen	Carbon	Boron	Arsenic
Silver	Chromium	Sulphur	Oxygen	Silicon	Antimony
Copper	Mercury	Phosphorus	Nitrogen	Germanium	Tellurium

COMPOUNDS – A **compound** is a pure substance made up of two or more elements combined chemically in a definite (or constant) proportion by mass (or by weight).

Water, Carbon dioxide and Sugar are some common compounds.

- Water is a compound made up of hydrogen and oxygen combined in the ratio 2 : 1 by volume and 1 : 8 by mass.

Hydrogen	+	Oxygen	=	Water
2 atom		1 atom		1 molecule
2 volume		1 volume		1 volume

Compounds are also known as **chemical compounds,** because these are formed due to chemical reactions.

- Carbon dioxide is a compound made up of carbon and oxygen in the ratio 1 : 2 by number and 3 : 8 by mass.

Carbon	+	Oxygen	=	Carbon dioxide
1 atom		2 atom		1 molecule

Self Assessment 1

1. Name the smallest particle of an element that takes part in a chemical reaction.
2. How many atoms does a molecule of water contain?
3. Name the compound containing oxygen and hydrogen.
4. Which form of matter occupies a definite space but has no definite shape?
5. Which of the following is a nonmetal? Silver, Sulphur, Silicon, Copper.

STRUCTURE OF AN ATOM

The modern model of an atom is based on the researches made mainly by Rutherford and Böhr. According to this model,

- The atom consists of a positively charged core called the **nucleus.**
- Almost the entire mass of an atom is concentrated in its nucleus.
- The nucleus consists of protons and neutrons.
- The nucleus of hydrogen atom, however, consists of only one proton.
- In an atom, the nucleus is surrounded by electrons moving around it in certain fixed orbits.
- An atom as a whole is electrically neutral.
- The number of protons inside the nucleus of an atom is equal to the number of electrons surrounding the nucleus.
- Electrons, protons and neutrons are called the **subatomic particles.**

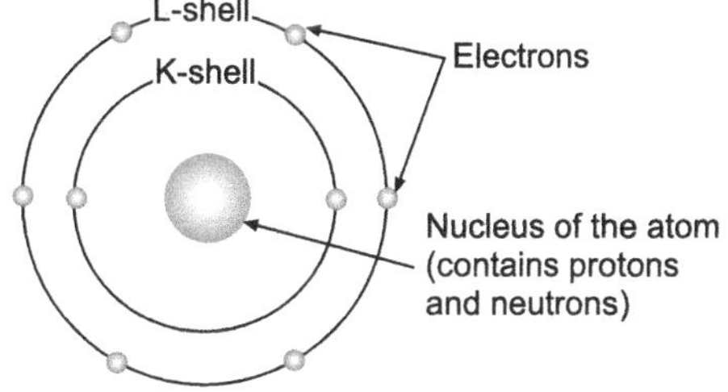

Fig. 1.1 Model of an atom
(electrons moving around the nucleus)

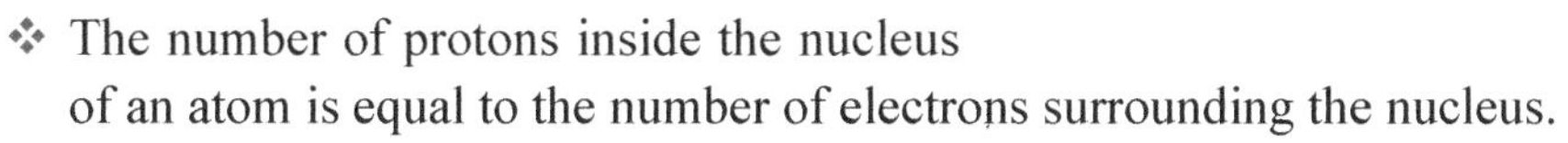

ELECTRON – The electron is a negatively charged subatomic particle. The mass of an electron is about $\frac{1}{1837}$ times that of a hydrogen atom. Thus, the mass of an electron is negligible.

> The negative charge possessed by an electron is considered as one unit of negative charge.

PROTON – The proton is a positively charged subatomic particle. The mass of a proton is almost equal to that of a hydrogen atom. Thus, a proton is 1837 times heavier than an electron.

NEUTRON – The neutron is a subatomic particle which does not carry any electrical charge. The mass of a neutron is almost equal to that of a proton. Thus, the mass of a neutron is taken as one atomic mass unit (or 1 amu or 1 u), where 1 u = 1.66×10^{-24} g.

> The positive charge possessed by a proton is considered as one unit of positive charge.

The symbol for atomic mass unit is **u**.

Particle	Symbol	Mass	Charge
Electron	e	Negligible	–1 unit
Proton	p	1 u	+1 unit
Neutron	n	1 u	0

ATOMIC NUMBER – Atomic number of an element is the number of protons present inside the nucleus of that element. It is denoted by *Z*.

Atomic number cannot have a fractional value.

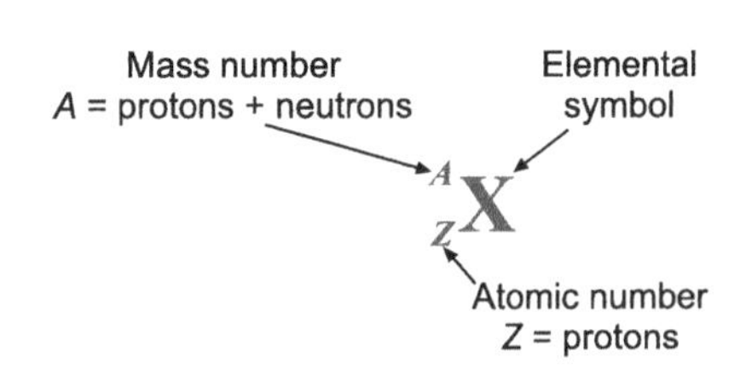

MASS NUMBER – Sum of the number of protons and neutrons inside the nucleus of an atom is called its mass number. It is denoted by *A*.

ATOMIC MASS – The average mass of an atom of an element in atomic mass units is called its atomic mass. Atomic mass is denoted by *A*.

The atomic mass of hydrogen is 1.008 atomic mass unit (or 1.008 u). Commonly, the term 'relative atomic mass (A_r)' is also used.

RELATIVE ATOMIC MASS

The relative atomic mass (A_r) of an element is defined as the average relative mass of its atom as compared with that of an atom of $^{12}_{6}C$ (carbon-12) taken as 12 atomic mass unit. **Relative atomic mass has no units.** It is a pure number.

Number of electrons, protons and neutrons in an atom of any element

Number of electrons (*e*) = Number of protons (*P*) = Atomic number (*Z*)

Mass number (*A*) = Number of protons (*P*) + Number of neutrons (*N*)

The atomic masses, number of electrons, protons and neutrons in the atoms of certain common elements is given below.

Number of Electrons, Protons and Neutrons in an Atom

Element		Atomic number (*Z*)	Mass number (*A*)	Number of		
Symbol	Name			Neutrons (*N*)	Protons (*P*)	Electrons (*e*)
$^{1}_{1}H$	Hydrogen	1	1	0	1	1
$^{4}_{2}He$	Helium	2	4	2	2	2
$^{12}_{6}C$	Carbon	6	12	6	6	6
$^{14}_{7}N$	Nitrogen	7	14	7	7	7
$^{16}_{8}O$	Oxygen	8	16	8	8	8
$^{23}_{11}Na$	Sodium	11	23	12	11	11
$^{35}_{17}Cl$	Chlorine	17	35	18	17	17

Pictorial representations of atomic structures of some common elements are shown below.

Hydrogen (H) (1)

Helium (He) (2)

Carbon (C) (6)

7N 7P

Nitrogen (N) (7)

Oxygen (O) (8)

Sodium (Na) (11)

Chlorine (Cl) (17)

Self Assessment 2

1. Name two electrically charged subatomic particles.
2. How much is the mass of an electron relative to that of a hydrogen atom?
3. The nucleus of an atom consists of two protons and two neutrons. What is its atomic number and mass number?
4. Does an atom consist of equal number of protons and electrons?
5. A carbon atom consists of six protons and six electrons. What is its atomic number?

THE CHARGED PARTICLES

IONS – The electrically charged particle that is formed whenever an atom loses or gains one or more electrons is called an **ion.**

When an atom of hydrogen loses one electron, it forms a hydrogen ion.

$$\underset{\text{hydrogen atom}}{H} \rightarrow \underset{\text{hydrogen ion}}{H^+} + \underset{\text{electron}}{e^-}$$

CATION – The cation is a positively charged ion. It is formed whenever an atom loses one or more electrons.

When a sodium atom (Na) loses an electron, it gives sodium ion (Na^+). Na^+ is a cation.

$$\underset{\text{sodium atom}}{Na} \xrightarrow{\text{loss of electrons}} \underset{\text{sodium ion, positively charged ion (cation)}}{Na^+} + \underset{\text{electron}}{e^-}$$

The charge on any ion is shown as a **superscript** to the symbol of the corresponding element.

ANION – The anion is a negatively charged ion. It is formed whenever an atom gains one or more electrons.

When a chlorine atom (Cl) gains an electron, it gives chloride ion (Cl^-). Cl^- is an anion.

$$\underset{\text{chlorine atom}}{Cl} + \underset{\text{electron}}{e^-} \xrightarrow{\text{gain of electron}} \underset{\text{chloride ion, negatively charged ion (anion)}}{Cl^-}$$

RADICAL – A molecule of any compound consists of two parts — one coming from the parent acid and the other from the parent base. These parts may contain a single atom or a group of atoms. Each of these parts is called the **radical.**

ACID RADICAL – The part of a molecule of any compound coming from the parent acid is called the acid radical.

BASIC RADICAL – The part of a molecule of any compound coming from the parent base is called the basic radical.

- The radicals may be made up of one or more atoms of the same or different elements.
- The atoms within a radical are bonded to each other by covalent bonds.
- A radical may carry negative or positive charge.

For example:

- NH_4^+ is a positive radical (or basic radical).
- Cl^- is a negative radical (or acid radical).

ISOTOPES

Atoms of the same element, having the same atomic number but different mass numbers are called isotopes of that element.

In other words, isotopes are the atoms of the same element which have the same number of protons but different number of neutrons inside their nuclei.

For example: consider **particle count** in the three isotopes of oxygen described below.

Particle count	$^{16}_{8}O$	$^{17}_{8}O$	$^{18}_{8}O$
No. of protons (P)	8	8	8
No. of neutrons (N)	8	9	10

FRACTIONAL ATOMIC MASSES

All naturally occurring elements often consist of a mixture of isotopes. Each isotope has different mass number. Therefore, the atomic mass of an element is the weighted mean of the mass numbers of different isotopes.

For example: Chlorine (Cl) has two isotopes – chlorine - 35 ($^{35}_{17}Cl$), chlorine - 37 ($^{37}_{17}Cl$). Their proportion in the chlorine are 75.4% and 24.6%, respectively. So,

$$\text{Atomic mass of chlorine} = \frac{(75.4 \times 35 + 24.6 \times 37)}{100}\ u$$

$$= \frac{(2639 + 910.2)}{100}\ u$$

$$= 35.49\ u$$

The atomic mass of chlorine is 35.49 atomic mass units (or 35.49 u).

Self Assessment 3

Fill in the blanks:

1. Cations are ______________ charged ions.
2. Negative ion is formed when an atom ______________ one or more electrons.
3. ______________ is an acid radical.
4. Basic radicals are ______________ charged.
5. When potassium atom loses one electron, the ion formed is ______________ .

REVIEW QUESTIONS

A. Very Short Answer Type Questions

1. Name the smallest particle of matter which can exist independently.
2. Which form of matter occupies a definite space and has definite shape?
3. Which of the following is/are metals?

   Hydrogen,Gold, Boron, Mercury, Ice, Carbon.
4. Name the three subatomic particles present in an atom.
5. Mention one difference between an electron and a proton.
6. Which of the following is heavier than the other? Proton, Electron.
7. Which number describes the sum of the numbers of protons and neutrons inside the nucleus of an atom?
8. What is the unit of atomic mass?
9. Which of the following is/are cation? Cl^-, Na^+, SO_4^{2-}.
10. Name one each of acid radical and basic radical.

B. Short Answer Type Questions

1. How is matter classified on the basis of its physical states? Which of them possesses definite shape and definite size?
2. What is a compound? Give its two examples.
3. How are protons, neutrons and electrons arranged in an atom?
4. Which ion carries positive charge? Give one example.
5. What is an acid radical? Give one example.

C. Long Answer Type Questions

1. What are elements? How are these classified? Give one example of each.
2. What are the subatomic particles? Name them. Which of them carry electrical charge?
3. Using the data given below indicate the species that represents
   (*a*) a cation (*b*) an anion (*c*) an atom?

Species	Number of protons	Number of neutrons	Number of electrons
A	9	10	11
B	17	18	17
C	6	6	6

4. Calculate the number of protons, neutrons and electrons in neutral atoms of the following isotopes.

   i. ${}^{1}_{1}H$, ${}^{2}_{1}H$ and ${}^{3}_{1}H$ **ii.** ${}^{235}_{92}U$ and ${}^{238}_{92}U$
5. The isotopic abundances of three isotopes of oxygen are:

   ${}^{16}_{8}O$ (99.76%), ${}^{17}_{8}O$ (0.04%), ${}^{18}_{8}O$ (0.20%)

   Calculate the atomic mass of oxygen.

6. Composition of three atomic species X, Y and Z are given as under

	X	Y	Z
No. of protons	12	12	12
No. of neutrons	12	13	14
No. of electrons	12	12	12

What is the relation between these species?

D. True or False Type Questions

Write 'T' for True or 'F' for False.

1. Matter is electrical in nature and negatively charged.
2. Atoms are always electrically charged.
3. Matter can be classified into elements and compounds.
4. Element is the simplest form of matter.
5. The mass of an electron is about 1840 times that of proton.
6. The mass number of an element is equal to the number of protons inside the nucleus of an atom of that element.
7. A neutron is formed by an electron and a proton combining together. Therefore, it is electrically neutral.
8. Anions are electrically neutral.
9. Acid radicals carry negative charge.
10. CO_3^{2-} is a basic radical.

E. Fill in the Blanks Type Questions

1. A compound is made up of ______________.
2. An electrically charged particle is known as a/an ______________.
3. The three states of matter are solid, ______________ and ______________.
4. An element is composed of ______________ of the same kind.
5. Letter *A* is used to denote the ______________ of an element.
6. The ______________ occupies a definite space but it has no definite shape.
7. A proton is a ______________ charged subatomic particle.
8. Elements are classified as metals, nonmetals and ______________.
9. Complete the following table.

Element	Symbol	Oxidation State	Ion type	Atomic mass
Calcium	....	+2	....	40
Sulphur	S	−2	....	....
....	Na	....	Cation	23
....	Be	+2	Cation	....
Chlorine	....	....	Anion	35.5
....	F	−1	....	19

10. Complete the following table.

Element	Atomic number	Atomic mass	Number of protons	Number of electrons	Number of neutrons
Silver	47	....	....	....	....
Cobalt	....	60	....	27	....
....	16	....	....	....	16
....	....	64	29	....	....
....	....	....	....	18	22

F. Match the Columns Type Questions

a.

	Column A		Column B
1	Ice	A.	Metal
2	Gold	B.	Cation
3	Helium	C.	Solid
4	Cs^+	D.	Gas
5	Water	E.	Compound

b.

	Column A		Column B
1	Metalloid	A.	Solid
2	S^{2-}	B.	Aluminium ion
3	Iron	C.	Basic radical
4	Al^{3+}	D.	Boron
5	Li^+	E.	Acid radical

G. Multiple Choice Type Questions

Choose the correct alternative.

1. Iron is an example of a
   (a) liquid ❑ (b) metal ❑ (c) compound ❑ (d) molecule ❑
2. Atomic number (Z) of carbon is
   (a) 1 ❑ (b) 12 ❑ (c) 6 ❑ (d) 4 ❑
3. Based on the physical state, matter can be classified into
   (a) gas ❑ (b) liquid ❑ (c) solid ❑ (d) all of these ❑
4. Gain of one or more electrons results in the formation of
   (a) cation ❑ (b) anion ❑ (c) both ❑ (d) none of these ❑
5. The type of charge carried by a neutron is
   (a) no charge ❑ (b) negative ❑ (c) positive ❑ (d) not certain ❑
6. The form of matter that is more stable at higher temperature is
   (a) liquid ❑ (b) gas ❑ (c) solid ❑ (d) all of these ❑
7. Which of the following is one of the subatomic particles present in an atom?
   (a) neutron ❑ (b) proton ❑ (c) electron ❑ (d) all of these ❑
8. One atomic mass unit is equivalent to the mass of
   (a) one electron ❑ (b) one proton ❑ (c) one neutron ❑ (d) none of these ❑
9. The nucleus of hydrogen atom is called
   (a) neutron ❑ (b) positron ❑ (c) proton ❑ (d) positive rays ❑
10. An atom consists of an equal number of
   (a) neutrons and protons ❑ (b) protons and electrons ❑
   (c) neutrons and electrons ❑ (d) protons and positrons ❑

ANSWERS

A. 1. Molecule 2. Solid 3. Gold, Mercury 4. Electron, Proton, Neutron
5. Electron carries negative charge and proton carries positive charge.
6. Proton 7. Mass number 8. Atomic mass unit 9. Na^+
10. Acid radical: Cl^- Basic radical: Na^+

D. 1. F 2. F 3. T 4. T 5. F 6. F 7. T 8. F 9. T 10. F

E. 1. Two or more elements 2. Ion 3. Liquid, Gas 4. Atoms 5. Atomic mass
6. Liquid 7. Positively 8. Metalloids
9. **Column-wise from left to right:** Ca, Cation; Anion, 32; Sodium, +1; Beryllium, 9; Cl, –1; Fluorine, Anion
10. **Column-wise from left to right:** 107, 47, 47, 60; 27, 27, 33; Sulphur, 32, 16, 16; Copper, 29, 29, 35; Argon, 18, 40, 18

F. a. 1 – C; 2 – A; 3 – D; 4 – B; 5 – E
b. 1 – D; 2 – E; 3 – A; 4 – B; 5 – C

G. 1. b 2. c 3. d 4. b 5. a 6. b 7. d 8. b 9. b 10. c

Four New Elements Discovered

Discovery of four new elements completes the 7th period of the Modern Periodic table. The International Union of Pure and Applied Chemistry has announced the names for these new elements.

All four elements were synthetically created in laboratories.

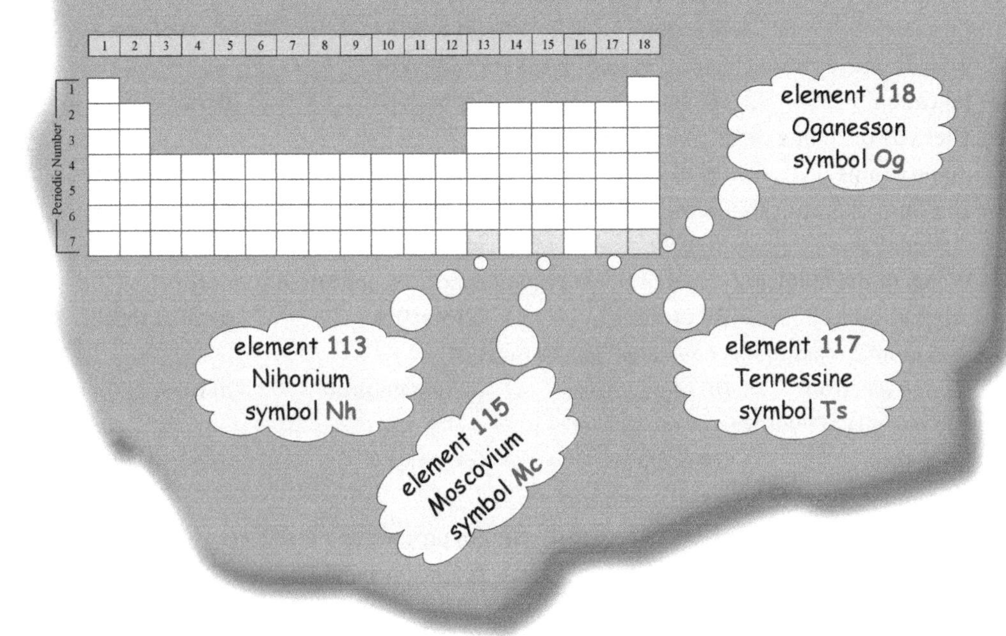

2

Symbols and Formulae

J. J. Berzelius introduced the modern system of denoting elements by their symbols.

Symbol – The symbol of an element is an abbreviation (short name) for its full name.

or

The symbol of an element is a shorthand notation for its full name.

Modern Atomic Symbols – According to the modern system of atomic symbols,

- The first letter of the Latin or Greek name of an element is taken as its symbol. Such a letter is written as a capital letter.
- If more than one element has the same first letter, their symbols consist of two letters – the first letter is followed by another letter which is not common. The first letter is written as a capital letter while the second letter as a small letter.

For example, the symbol of **Hydrogen** is **H** and that of **Helium** is **He.** The symbol of **Calcium** is **Ca** and that of **Chromium** is **Cr.**

The common names of some typical elements, their symbols and the source from where these symbols are derived, are given below:

Atomic Symbols for Some Typical Elements

Carbon	Carbonium	(Latin)	C	Calcium	Calx	(Latin)	Ca
Hydrogen	Hydrogenium	(Latin)	H	Copper	Cuprum	(Latin)	Cu
Nitrogen	Nitrogenium	(Latin)	N	Iron	Ferrum	(Latin)	Fe
Oxygen	Oxygenium	(Latin)	O	Silicon	Silex	(Latin)	Si
Potassium	Kalium	(Latin)	K	Sodium	Natrium	(Latin)	Na

Self Assessment 1

1. The modern system of denoting elements by their symbols was given by __________
2. The shorthand notation for the full name of an element is known as its __________
3. Give the atomic symbols of
   (a) Iodine __________ (b) Gold __________ (c) Magnesium __________
   (d) Fluorine __________ (e) Boron __________
4. The symbol of sodium is Na but the letters Na are not the starting letters in sodium. From where is the symbol of sodium derived?

CHEMICAL FORMULA

Each chemical compound is known by a specific **name,** and is denoted by its **chemical formula.** There are two types of chemical formulae:

- Molecular formula
- Empirical formula

MOLECULAR FORMULA – A shorthand notation of a molecule in terms of symbols and the number of atoms of various elements present in it is called its **molecular formula.**

or

The symbolic representation of a molecule of any substance describing the actual number of various atoms present in it is called its **molecular formula.**

For example:

- The molecular formula of water is H_2O. Thus, one molecule of water contains two atoms of hydrogen (H) and one atom of oxygen (O).
- The molecular formula of carbon dioxide is CO_2. Thus, one molecule of CO_2 contains one atom of carbon (C) and two atoms of oxygen (O).

EMPIRICAL FORMULA – The simplest formula of a substance which gives the lowest whole number ratio between the number of atoms of different elements present in the substance is called its **empirical formula.**

Writing an empirical formula can be illustrated in the following table.

Illustrating the Writing of Empirical Formula

Compound	Molecular formula	Ratio of atoms in the molecule	Lowest ratio	Empirical formula
Sodium chloride	NaCl	1 : 1	1 : 1	NaCl
Benzene	C_6H_6	6 : 6	1 : 1	CH
Ethene	C_2H_4	2 : 4	1 : 2	CH_2

MOLECULAR FORMULA OF AN ELEMENT

Follow the steps given below:

- Write the atomic symbol of the element.
- Write the number of atoms present in its molecule as a subscript on the right-hand side of the symbol.

For example, a molecule of hydrogen contains two hydrogen atoms. So, the molecular formula of hydrogen is H_2.

A molecule of an element constitutes atoms of the same kind

Symbol of hydrogen ⟵ H_2 ⟶ Subscript showing the number of atoms of hydrogen in its molecule

Molecular formulae of some common elements are given below:

Molecular Formulae of Some Common Elements

Element		No. of atoms in one molecule	Molecular formula
Name	Symbol		
Helium	He	1	He
Oxygen	O	2	O_2
Nitrogen	N	2	N_2
Chlorine	Cl	2	Cl_2
Bromine	Br	2	Br_2
Ozone*	O	3	O_3
Phosphorus	P	4	P_4
Sulphur	S	8	S_8

*Ozone is an allotrope of oxygen.

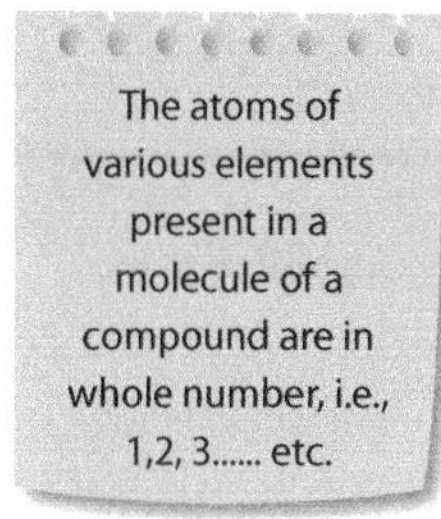

MOLECULAR FORMULA OF A COMPOUND

Follow the steps given below:

- Write the atomic symbols of the constituent elements.
- Write the number of atoms of each element present in the molecule as a subscript to the respective symbol. The subscript 1 is not written.

For example, a molecule of hydrogen chloride contains one atom each of hydrogen and chlorine. So,

$$\text{Molecular formula of hydrogen chloride} = H_1Cl_1 \Rightarrow HCl$$

- A molecule of water contains two atoms of hydrogen and one atom of oxygen. So,

$$\text{Molecular formula of water} = H_2O_1 \Rightarrow H_2O$$

- A molecule of ammonium chloride contains one atom of nitrogen, four atoms of hydrogen and one atom of chlorine. So,

$$\text{Molecular formula of ammonium chloride} = N_1H_4Cl_1 \Rightarrow NH_4Cl$$

ATOMICITY OF A MOLECULE – The number of atoms present in a molecule of any element or a compound is called its **atomicity.**

Atomicity of Some Elements at Room Temperature

Element	Molecular formula	No. of atoms in a molecule	Atomicity	Element	Molecular formula	No. of atoms in a molecule	Atomicity
Helium	He	1	1	Chlorine	Cl_2	2	2
Hydrogen	H_2	2	2	Ozone	O_3	3	3
Oxygen	O_2	2	2	Phosphorus	P_4	4	4
Nitrogen	N_2	2	2	Sulphur	S_8	8	8

The molecular formulae of certain common compounds and their atomicities are listed in Table given below:

Molecular Formulae and Atomicity of Some Common Compounds

Name of compound	Constituent elements of compound		No. of atoms of each element in a molecule of the compound	Molecular formula of the compound		Atomicity of the molecule
Carbon dioxide	Carbon	(C)	1	C_1O_2	CO_2	3
	Oxygen	(O)	2			
Sulphuric acid	Hydrogen	(H)	2	$H_2S_1O_4$	H_2SO_4	7
	Sulphur	(S)	1			
	Oxygen	(O)	4			
Sugar	Carbon	(C)	12	$C_{12}H_{22}O_{11}$	$C_{12}H_{22}O_{11}$	45
	Hydrogen	(H)	22			
	Oxygen	(O)	11			
Potassium permanganate	Potassium	(K)	1	$K_1Mn_1O_4$	$KMnO_4$	6
	Manganese	(Mn)	1			
	Oxygen	(O)	4			
Ammonia	Nitrogen	(N)	1	N_1H_3	NH_3	4
	Hydrogen	(H)	3			
Nitric acid	Hydrogen	(H)	1	$H_1N_1O_3$	HNO_3	5
	Nitrogen	(N)	1			
	Oxygen	(O)	3			
Phosphorus pentoxide	Phosphorus	(P)	2	P_2O_5	P_2O_5	7
	Oxygen	(O)	5			
Mercuric chloride	Mercury	(Hg)	1	Hg_1Cl_2	$HgCl_2$	3
	Chlorine	(Cl)	2			
Aluminium bromide	Aluminium	(Al)	1	Al_1Br_3	$AlBr_3$	4
	Bromine	(Br)	3			
Borax	Sodium	(Na)	2	$Na_2B_4O_7$	$Na_2B_4O_7$	13
	Boron	(B)	4			
	Oxygen	(O)	7			

Self Assessment 2

Fill in the blanks:

1. A molecule of an element constitutes ____________ of the same kind.
2. The empirical formula of a compound in which carbon and hydrogen are present in the ratio 1 : 1.67 is ____________.
3. Specific name for each chemical compound is denoted by its ____________.
4. Atomicity of H_3PO_4 molecule is ____________.
5. The molecular formula of an element X having atomicity of 4 is ____________.

REVIEW QUESTIONS

A. Very Short Answer Type Questions

1. Who introduced the modern system of atomic symbols?
2. Name the element whose symbol is Cu.
3. What describes/denotes the specific name of a chemical compound?
4. What does the molecular formula CO_2 denote?
5. Write the molecular formula of ozone.
6. What is the atomicity of the molecule NH_3?
7. Name an element whose atomicity is one.
8. From which name the symbol of iron (Fe) is derived?
9. Write the molecular formula of a triatomic molecule.
10. What is the molecular formula of an element X having atomicity of 7?

B. Short Answer Type Questions

1. What is meant by the symbol of an element?
2. Define the molecular formula of a compound.
3. The molecular formula of a compound is N_2O_4. What is its empirical formula?
4. The molecule of a compound contains two atoms of hydrogen and one atom of sulphur. What is its molecular formula?
5. What are the atomicities of the following molecules?
   P_4, O_2, He, P_2O_5 and NH_3.

C. Long Answer Type Questions

1. Define the atomic symbol. How are the atomic symbols of elements written?
2. How does the empirical formula of a compound differ from its molecular formula?
3. How is the molecular formula of a compound written?
4. What is meant by the term 'atomicity of a molecule'?
5. What are the atomicities of the following compounds?
   HNO_3, $C_6H_{12}O_6$, $KMnO_4$, $CaCO_3$ and $Na_2B_4O_7$.

D. True or False Type Questions

Write 'T' for True or 'F' for False.

1. Symbol for potassium is P.
2. Sodium is represented by the symbol Na.
3. Mercury is represented by the symbol Hg.
4. Ag is the symbol for metal silver.
5. It is possible to have an atom of a compound.
6. Formula stands for one atom of an element.
7. N_2O_4 describes the empirical formula of dinitrogen tetraoxide.
8. CO is an empirical formula for carbon dioxide gas.
9. Atomicity of water molecule is two.
10. A molecule of sodium oxide has atomicity equal to four.

E. Fill in the Blanks Type Questions

1. Write the symbol of each element in the space provided.

Element	Symbol	Element	Symbol	Element	Symbol	Element	Symbol
Hydrogen	....	Bromine	....	Nitrogen	....	Lithium	....
Helium	....	Fluorine	....	Aluminium	....	Sodium	....
Neon	....	Iodine	....	Boron	....	Potassium	....
Argon	....	Beryllium	....	Calcium	....	Cobalt	....
Tin	....	Magnesium	....	Copper	....	Iron	....

2. Write the names of the elements in the space provided.

Symbol	Element	Symbol	Element	Symbol	Element	Symbol	Element
B	....	K	....	Br	....	Cl	....
C	....	N	....	Fe	....	Ca	....
F	....	O	....	Mg	....	Cu	....
H	....	P	....	Ni	....	Co	....
I	....	S	....	Zn	....	Cr	....

3. Fill in the blanks in the following table.

	Molecular formula	Constituent elements	Atomicity	Name of the compound
a.	$KMnO_4$	K, Mn, O	....	....
b.	....	....	2	Potassium chloride
c.	....	Na, P, O	8	....
d.	P_2O_5	....	....	Phosphorus pentoxide
e.	....	Al, Cl	4	....
f.	$CaCO_3$	....	....	Calcium carbonate
g.	....	Ca, N, O	....	Calcium nitrate
h.	NaOH	....	3	....
i.	Cu_2O	....	....	Copper oxide
j.	....	Al, S, O	17	....
k.	$Ba(OH)_2$	Ba, O, H	....	....
l.	....	....	6	Sodium sulphite

F. Match the Columns Type Questions

a.

	Column A		Column B
1	Copper	A.	CH_2O
2	Oxygen	B.	S_8
3	Empirical formula	C.	K
4	Molecular formula	D.	Cu
5	Potassium	E.	O_2

b.

	Column A		Column B
1	Barium	A.	Cr
2	Krypton	B.	B
3	Boron	C.	Au
4	Chromium	D.	Kr
5	Gold	E.	Ba

G. Multiple Choice Type Questions

Choose the correct alternative.

1. In writing the molecular formulae of elements, the number (subscript) on the right hand of the symbol denotes the
   (a) charge on the element ☐ (b) number of atoms in a molecule ☐
   (c) number of electrons in a molecule ☐ (d) number of protons in a molecule ☐
2. Silver is represented by the symbol
   (a) S ☐ (b) Si ☐ (c) Sr ☐ (d) none of these ☐
3. The molecular formula for carbon dioxide is
   (a) CO_2 ☐ (b) CO ☐ (c) CO_3 ☐ (d) none of these ☐
4. The symbol for metal iron is
   (a) I ☐ (b) Ir ☐ (c) Iro ☐ (d) none of these ☐
5. The correct symbol for copper is
   (a) C ☐ (b) Co ☐ (c) Cu ☐ (d) Cop ☐
6. Tin is represented by the symbol
   (a) T ☐ (b) Ti ☐ (c) Si ☐ (d) Sn ☐
7. Which of the following represents the symbol for potassium?
   (a) Pot ☐ (b) P ☐ (c) Po ☐ (d) none of these ☐
8. The atomicity of a CCl_4 molecule is
   (a) 6 ☐ (b) 5 ☐ (c) 2 ☐ (d) 1 ☐
9. The symbol of the noble gas argon is
   (a) Arg ☐ (b) Ar ☐ (c) Ag ☐ (d) A ☐
10. Atomicity of a molecule of ozone is
   (a) 1 ☐ (b) 2 ☐ (c) 3 ☐ (d) 4 ☐

ANSWERS

A. 1. J. J. Berzelius 2. Copper 3. Chemical composition 4. Carbon dioxide
5. O_3 6. Four 7. Helium 8. Ferrum 9. CO_2 10. X_7

D. 1. F 2. T 3. T 4. T 5. F 6. F 7. F 8. F 9. F 10. F

E. 1. **Column-wise from left to right:**
H, He, Ne, Ar, Sn; Br, F, I, Be, Mg; N, Al, B, Ca, Cu; Li, Na, K, Co, Fe

2.

Boron	Potassium	Bromine	Chlorine
Carbon	Nitrogen	Iron	Calcium
Fluorine	Oxygen	Magnesium	Copper
Hydrogen	Phosphorus	Nickel	Cobalt
Iodine	Sulphur	Zinc	Chromium

3. (a) 6, Potassium permanganate (b) KCl; K, Cl
(c) Na_3PO_4; Sodium phosphate (d) P, O; 7
(e) $AlCl_3$; Aluminium chloride (f) Ca, C, O; 5
(g) $Ca(NO_3)_2$; 9 (h) Na, O, H; Sodium hydroxide
(i) Cu, O; 3 (j) $Al_2(SO_4)_3$; Aluminium sulphate
(k) 5; Barium hydroxide (l) Na_2SO_3; Na, S, O

F. a. 1 – D; 2 – E; 3 – A; 4 – B; 5 – C
b. 1 – E; 2 – D; 3 – B; 4 – A; 5 – C

G. 1. b 2. d 3. a 4. d 5. c 6. d 7. d 8. b 9. b 10. c

Valency and Formulae

VALENCE ELECTRONS – The electrons present in the outermost shell in the atom of an element are called its **valence electrons.**

The outermost shell is also termed as valence shell.

Number of Valence Electrons in the Atoms of Some Elements

Element	Electronic configuration K	L	M	N	No. of electrons in the outermost shell	No. of valence electrons
H	1				1	1
C	2	4			4	4
O	2	6			6	6
Cl	2	8	7		7	7
K	2	8	8	1	1	1

VALENCY – The number of hydrogen atoms or chlorine atoms or double the number of oxygen atoms, which combine with one atom of the element, is called its **valency.** or

The number of electrons lost, gained or shared by its atom with some other atom to complete its octet during any chemical reaction is called its **valency**.

For example,

- In hydrogen chloride (HCl), one atom of chlorine (Cl) is combined with one atom of hydrogen (H). Therefore,

  Valency of chlorine in hydrogen chloride (HCl) = 1

- In ammonia (NH_3), one atom of nitrogen combines with three atoms of hydrogen (H). Therefore,

  Valency of nitrogen (N) in ammonia (NH_3) = 3

- In methane (CH_4), one atom of carbon is combined with four atoms of hydrogen. Therefore,

  Valency of carbon in methane (CH_4) = 4

In sodium chloride (NaCl), one atom of chlorine is combined with one atom of sodium (Na). Therefore,

Valency of chlorine in sodium chloride (NaCl) = 1

VARIABLE VALENCY – Certain elements show more than one valencies. Such elements are said to have **variable valency.**

The name of the ion having **lower valency** ends with **-ous,** while that of **higher valency** ends with **-ic.**

For example, Copper exists as Cu^{2+} and Cu^{+} and Iron exists as Fe^{3+} and Fe^{2+}.

THE STOCK NOTATION

According to the **Stock notation,** the valency of a metal is indicated by a **Roman numeral** enclosed in parentheses and written just after the symbol/name of the metal.

Illustrating the Variable Valency

Metal	Valency	Name of the compound	Constituent metal ion	Stock notation	Formula of the compound
Copper	1 +	Cuprous oxide	Cu^{+}	Cu(I)	Cu_2O
	2 +	Cupric oxide	Cu^{2+}	Cu(II)	CuO
Iron	2 +	Ferrous oxide	Fe^{2+}	Fe(II)	FeO
	3 +	Ferric oxide	Fe^{3+}	Fe(III)	Fe_2O_3

ELECTRONIC CONCEPT OF VALENCY

According to the **Electronic concept of valency,** the number of electrons which an atom loses, gains or shares with other atoms to attain the nearest noble gas configuration is termed as its **valency.**

There are two types of valencies.

- Electrovalency
- Covalency

ELECTROVALENCY – The number of electrons **lost** or **gained** by an atom of an element during the formation of an electrovalent bond is termed as its **electrovalency.**

- The element which **gives up** electrons to form positive ions have **positive valency.**
- The element which **accepts** electrons to form negative ions have **negative valency.**

For example, during the formation of sodium chloride (NaCl), the sodium atom loses one electron and the chlorine atom gains one electron.

Na	→	Na^{+}	+	e^{-}
sodium atom (2, 8, 1)		sodium ion (2, 8)		one electron

Cl	+	e^{-}	→	Cl^{-}
(2, 8, 7) chlorine atom		one electron		chloride ion (2, 8, 8)

Thus,

Electrovalency of sodium in sodium chloride = 1 +

Electrovalency of chlorine in sodium chloride = 1 –

- The ions or radicals which carry one unit of charge are called **monovalent.**
- Those carrying two units of charge are called **divalent** or **bivalent.**
- Those carrying three units of charge are called **trivalent.**
- Those carrying four units of charge are called **tetravalent.**

COVALENCY – The number of electrons contributed by an atom towards mutual sharing during the formation of a covalent bond is called **covalency.**

For example,

- Covalency of hydrogen in H_2 (hydrogen molecule) is **one.**

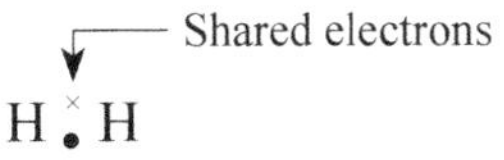

Shared electrons

H ×• H

Each H atom contributes one electron. So,

Covalency of H in H_2 = 1

- Covalency of hydrogen in NH_3 (ammonia molecule) is **one** and that of nitrogen is **three.**

H
×•
H ×• N
×•
H

The N atom contributes three electrons and the H atoms contribute one each to form NH_3. So,

Covalency of N in NH_3 = 3

Covalency of H in NH_3 = 1

- Covalency of carbon in methane (CH_4) is **four.**

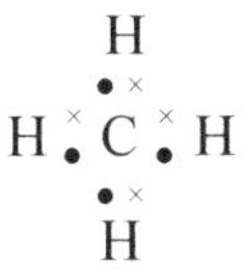

H
•×
H ×• C ×• H
•×
H

C atom contributes four electrons and the H atoms contribute one each to form CH_4. So,

Covalency of C in CH_4 = 4

Covalency of H in CH_4 = 1

Names, Formulae and Valencies of Some Common Ions/Radicals

Cation / Basic radical			Anion / Acid radical		
Name	Symbol	Valency	Name	Symbol	Valency
		MONOVALENT			
Hydrogen	H^+	1 +	Acetate	CH_3COO^-	1–
Sodium	Na^+	1 +	Hydrogen carbonate	HCO_3^-	1–
Potassium	K^+	1 +	Chloride	Cl^-	1–
Silver(I)	Ag^+	1 +	Bromide	Br^-	1–
Ammonium	NH_4^+	1 +	Hydroxide	OH^-	1–
Copper(I)	Cu^+	1 +	Nitrate	NO_3^-	1–
		DIVALENT			
Calcium	Ca^{2+}	2 +	Carbonate	CO_3^{2-}	2–
Copper(II)	Cu^{2+}	2 +	Chromate	CrO_4^{2-}	2–
Magnesium	Mg^{2+}	2+	Dichromate	$Cr_2O_7^{2-}$	2–
Nickel	Ni^{2+}	2 +	Oxide	O^{2-}	2–
Zinc	Zn^{2+}	2 +	Thiosulphate	$S_2O_3^{2-}$	2–
Lead(II)	Pb^{2+}	2 +	Sulphate	SO_4^{2-}	2–
Mercury(II)	Hg^{2+}	2 +	Sulphide	S^{2-}	2–
		TRIVALENT			
Aluminium	Al^{3+}	3 +	Borate	BO_3^{3-}	3–
Chromium	Cr^{3+}	3 +	Nitride	N^{3-}	3–
Iron(III)	Fe^{3+}	3 +	Phosphate	PO_4^{3-}	3–
Gold(III)	Au^{3+}	3 +			
		TETRAVALENT			
Lead(IV)	Pb^{4+}	4 +	Ferrocyanide	$[Fe(CN)_6]^{4-}$	4–
Tin(IV)	Sn^{4+}	4 +	Silicate	SiO_4^{4-}	4–
Platinum(IV)	Pt^{4+}	4 +			

MOLECULAR FORMULAE OF MOLECULAR COMPOUNDS

The molecular formula of a molecular compound is written as follows:

Step 1. Write the symbols of the constituent elements side-by-side, in such a way that the less electronegative element is on the left and the more electronegative element is on the right.

Step 2. Write their valency numbers over the symbols (as superscripts).

Step 3. Criss-cross the valency numbers to write as subscripts to the symbols.

Step 4. Divide these numbers (written as subscripts) by a common factor, if needed.

This method of writing the molecular formula of a molecular compound is illustrated through the following examples:

WRITING THE MOLECULAR FORMULA OF WATER

Water contains hydrogen and oxygen. Oxygen is more electronegative than hydrogen. The molecular formula of water can be written as follows:

Step 1. Writing the symbols of hydrogen and oxygen side by side

Step 2. Writing the valencies of hydrogen and oxygen on their symbols

Step 3. Criss-crossing the valencies of hydrogen and oxygen

Step 4. Deleting the subscript 1 and writing the molecular formula.

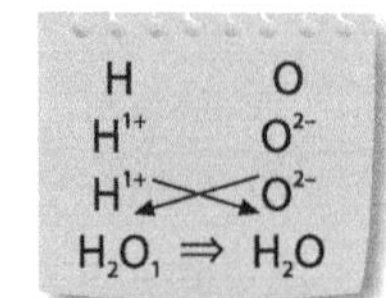

Thus, the molecular formula of water is H_2O.

WRITING THE MOLECULAR FORMULA OF HYDROGEN CHLORIDE

Hydrogen chloride contains hydrogen and chlorine. Chlorine is more electronegative than hydrogen. The molecular formula of hydrogen chloride can be written as follows:

Step 1. Writing the symbols of hydrogen and chlorine

Step 2. Writing the valencies of hydrogen and chlorine on their symbols

Step 3. Criss-crossing the valencies of hydrogen and chlorine

Step 4. Deleting the subscript 1 and writing the molecular formula.

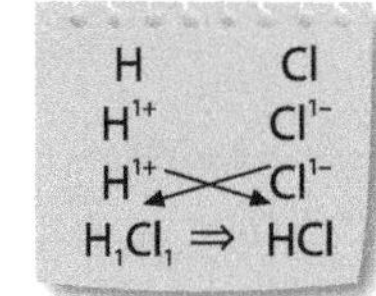

Thus, the molecular formula of hydrogen chloride is HCl.

WRITING THE MOLECULAR FORMULA OF HYDROGEN SULPHIDE

Hydrogen sulphide contains hydrogen and sulphur. Sulphur is more electronegative than hydrogen. Then, by following the steps as above, one can write

Step 1. Writing the symbols of hydrogen and sulphur

Step 2. Writing the valencies of hydrogen and sulphur on their symbols

Step 3. Criss-crossing the valencies of hydrogen and sulphur

Step 4. Deleting the subscript 1 and writing the molecular formula.

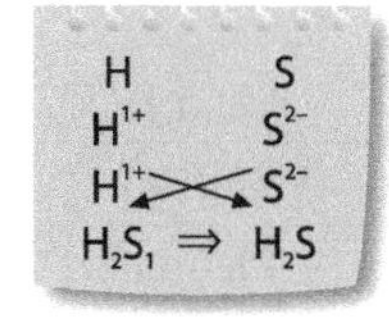

Thus, the molecular formula of hydrogen sulphide is H_2S.

WRITING THE MOLECULAR FORMULA OF CARBON TETRACHLORIDE

Carbon tetrachloride contains carbon and chlorine. Chlorine is more electronegative than carbon. Then, by following the steps as above, one can write

Step 1. Writing the symbols of carbon and chlorine

Step 2. Writing the valencies of carbon and chlorine on their symbols

Step 3. Criss-crossing the valencies of carbon and chlorine

Step 4. Deleting the subscript 1 and writing the molecular formula.

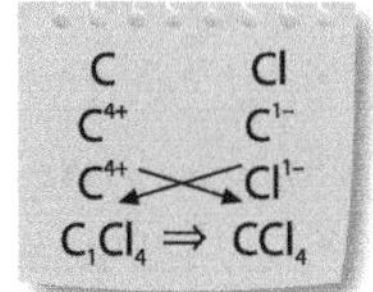

Thus, the molecular formula of carbon tetrachloride is CCl_4.

Self Assessment 1

1. How many valence electrons are there in an atom of oxygen?
2. What is the valency of nitrogen (N) in ammonia (NH_3)?
3. Copper occurs in Cu^+ and Cu^{2+} states. Describe these ions in terms of Stock's notation.
4. Write the molecular formula of a compound that contains two hydrogen atoms and one sulphur atom.
5. Find the valency of M in the compound M_2O.

FORMULAE OF IONIC COMPOUNDS

The ionic (or electrovalent) compounds contain ions. These ions are arranged in such a way that a cation has an anion as its nearest neighbour and vice-versa. There is no discrete molecule of an ionic compound. Therefore, **it is not correct to assign a molecular formula to an ionic compound.** Instead, an ionic compound is described by a formula which describes a **simple atomic ratio** of the elements present in it.

The formula which describes the simplest atomic ratio of the elements present in a compound is called its **stoichiometric formula** or **chemical formula.**

Sodium chloride is an ionic compound in which sodium and chlorine are present in 1:1 ratio. So, the stoichiometric or chemical formula of sodium chloride is Na^+Cl^- (or NaCl).

WRITING THE CHEMICAL FORMULA OF AN IONIC COMPOUND

Step 1. Write symbol of the **cation** showing its charge number at the right top (as superscript).

Step 2. Write symbol of the **anion** showing its charge number at the right top (as superscript).

Step 3. If a compound contains polyatomic ion, then the formula of the polyatomic ion is enclosed within brackets before writing their valencies.

Step 4. Write the charge number (valency) of the cation at the bottom-right of the anion (as subscript) and the charge number of the anion at the bottom-right of the cation (as subscript). Thus, the symbol of cation is subscripted with the charge number of the anion and the anion is subscripted with the charge number of the cation. This is called the **criss-crossing of valencies.**

Step 5. If these subscripts are 1, these are not written in the final formula. Otherwise, these subscripts are reduced to the lowest possible integers by dividing each by the highest common factor.

This method of writing the chemical formulae for ionic compounds is illustrated below:

WRITING THE CHEMICAL FORMULA OF ALUMINIUM SULPHATE

Aluminium sulphate contains Al^{3+} and SO_4^{2-} ions. The formula for aluminium sulphate can be obtained as follows:

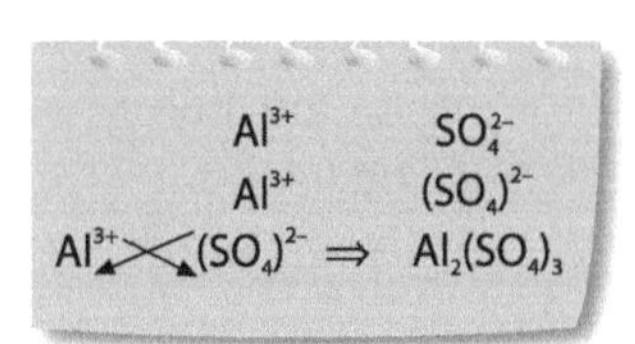

Step 1. Writing the two ions side-by-side along with their valencies

Step 2. Enclosing the (SO_4^{2-}) ion within brackets placing its valency outside the bracket on RHS

Step 3. Criss-crossing of the valencies

Step 4. There is no common factor between 2 and 3.

Therefore, the formula of aluminium sulphate is $Al_2(SO_4)_3$.

WRITING THE CHEMICAL FORMULA OF SODIUM HYDROGENCARBONATE

Sodium hydrogencarbonate (or sodium bicarbonate) contains Na^+ and HCO_3^- ions. The formula of sodium hydrogencarbonate is obtained as follows:

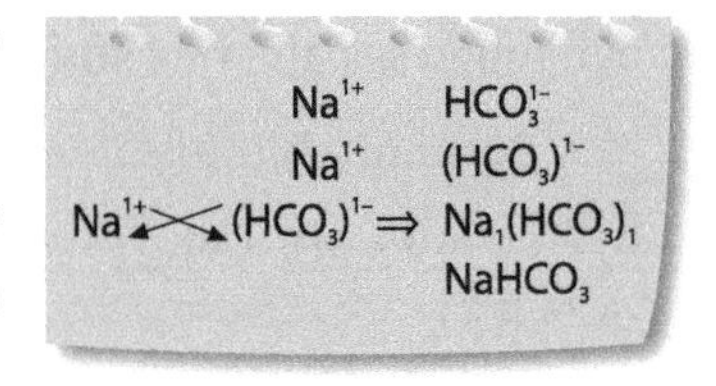

Step 1. Writing the symbols of the two ions side-by-side along with their valencies.

Step 2. Enclosing the polyatomic HCO_3^{1-} ion within the brackets placing its valency outside the bracket on RHS

Step 3. Criss-crossing of the valencies

Step 4. Deleting the subscripts 1 and writing the formula.

Therefore, the formula of sodium hydrogencarbonate is $NaHCO_3$.

WRITING THE CHEMICAL FORMULA OF BARIUM CARBONATE

Barium carbonate contains Ba^{2+} and CO_3^{2-} ions. The formula of barium carbonate can be obtained as follows:

Step 1. Writing the two ions, Ba^{2+} and CO_3^{2-} side-by-side along with their valencies

Step 2. Enclosing the CO_3^{2-} ion within brackets and placing its valency outside the bracket on RHS

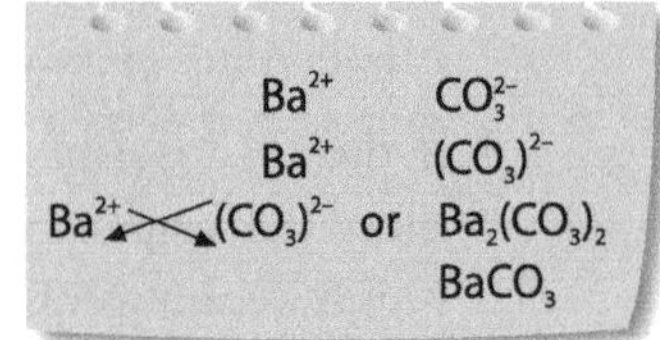

Step 3. Criss-crossing of the valencies

Step 4. Cancelling out the common factor and writing the formula.

Therefore, the formula of barium carbonate is $BaCO_3$.

The scheme of writing chemical formulae of some typical compounds is illustrated in the following table.

Chemical Formulae of Some Typical Compounds

Name of the compound	Positive ion (cation)			Negative ion (anion)			Chemical formula
	Name	Formula or Symbol	Valency	Name	Formula or Symbol	Valency	
Hydrogen chloride	Hydrogen	H	1	Chloride	Cl	1	$H^1 \times Cl^1 \rightarrow H_1Cl_1 \rightarrow HCl$
Hydrogen sulphide	Hydrogen	H	1	Sulphide	S	2	$H^1 \times S^2 \rightarrow H_2S_1 \rightarrow H_2S$
Sulphuric acid (Hydrogen sulphate)	Hydrogen	H	1	Sulphate	SO_4	2	$H^1 \times (SO_4)^2 \rightarrow H_2(SO_4)_1 \rightarrow H_2SO_4$
Sodium nitrate	Sodium	Na	1	Nitrate	NO_3	1	$Na^1 \times NO_3^1 \rightarrow Na_1(NO_3)_1 \rightarrow NaNO_3$
Aluminium phosphate	Aluminium	Al	3	Phosphate	PO_4	3	$Al^3 \times (PO_4)^3 \rightarrow Al_3(PO_4)_3 \rightarrow AlPO_4$
Aluminium sulphate	Aluminium	Al	3	Sulphate	SO_4	2	$Al^3 \times (SO_4)^2 \rightarrow Al_2(SO_4)_3$
Ferrous sulphate	Ferrous	Fe	2	Sulphate	SO_4	2	$Fe^2 \times (SO_4)^2 \rightarrow Fe_2(SO_4)_2 \rightarrow FeSO_4$
Ferric sulphate	Ferric	Fe	3	Sulphate	SO_4	2	$Fe^3 \times (SO_4)^2 \rightarrow Fe_2(SO_4)_3$
Potassium dichromate	Potassium	K	1	Dichromate	Cr_2O_7	2	$K^1 \times (Cr_2O_7)^2 \rightarrow K_2(Cr_2O_7)_1 \rightarrow K_2Cr_2O_7$
Magnesium nitrate	Magnesium	Mg	2	Nitrate	NO_3	1	$Mg^2 \times (NO_3)^1 \rightarrow Mg(NO_3)_2$
Silver chromate	Silver	Ag	1	Chromate	CrO_4	2	$Ag^1 \times (CrO_4)^2 \rightarrow Ag_2CrO_4$
Barium carbonate	Barium	Ba	2	Carbonate	CO_3	2	$Ba^2 \times (CO_3)^2 \rightarrow Ba_2(CO_3)_2 \rightarrow BaCO_3$
Potassium permanganate	Potassium	K	1	Permanganate	MnO_4	1	$K^1 \times (MnO_4)^1 \rightarrow KMnO_4$
Calcium hydroxide	Calcium	Ca	2	Hydroxide	OH	1	$Ca^2 \times (OH)^1 \rightarrow Ca(OH)_2$
Aluminium oxide	Aluminium	Al	3	Oxide	O	2	$Al^3 \times O^2 \rightarrow Al_2O_3$
Magnesium phosphate	Magnesium	Mg	2	Phosphate	PO_4	3	$Mg^2 \times (PO_4)^3 \rightarrow Mg_3(PO_4)_2$
Ammonium sulphite	Ammonium	NH_4	1	Sulphite	SO_3	2	$(NH_4)^1 \times (SO_3)^2 \rightarrow (NH_4)_2SO_3$
Zinc phosphate	Zinc	Zn	2	Phosphate	PO_4	3	$Zn^2 \times (PO_4)^3 \rightarrow Zn_3(PO_4)_2$

Note: The signs indicating the nature of charge, that is, positive (+) or negative (–), are omitted for the sake of convenience.

SOME TYPICAL EXAMPLES

Example 1: Write the formulae of the compounds containing the following ions:

(a) Cr^{3+} and F^-

(b) Hg^{2+} and S^{2-}

(c) Pb^{2+} and PO_4^{3-}

Solution: The formulae of the compounds formed by the given pairs of ions are given below:

	Ions	Formula	
a.	Cr^{3+} and F^{-}	$Cr^{3+} \times F^{1-}$	$Cr_1F_3 \Rightarrow CrF_3$
b.	Hg^{2+} and S^{2-}	$Hg^{2+} \times S^{2-}$	$Hg_2S_2 \Rightarrow HgS$
c.	Pb^{2+} and PO_4^{3-}	$Pb^{2+} \times (PO_4)^{3-}$	$Pb_3(PO_4)_2$

Example 2: Write the formulae of the following compounds:

(a) Ammonium carbonate

(b) Barium sulphate

(c) Calcium phosphate

Solution: The formulae of the given compounds are given below:

	Compound	Ions	Formula	
a.	Ammonium carbonate	$NH_4^+ \times CO_3^{2-}$	$(NH_4)_2(CO_3)_1$	$\Rightarrow (NH_4)_2CO_3$
b.	Barium sulphate	$Ba^{2+} \times SO_4^{2-}$	$Ba_2(SO_4)_2$	$\Rightarrow BaSO_4$
c.	Calcium phosphate	$Ca^{2+} \times PO_4^{3-}$	$Ca_3(PO_4)_2$	$\Rightarrow Ca_3(PO_4)_2$

Example 3: Write the chemical formulae of the following compounds:

(a) Magnesium chloride (b) Calcium oxide

(c) Copper nitrate (d) Aluminium chloride

(e) Calcium carbonate

Solution:

	Compound	Ions	Formula	
a.	Magnesium chloride	$Mg^{2+} \times Cl^{1-}$	Mg_1Cl_2	$\Rightarrow MgCl_2$
b.	Calcium oxide	$Ca^{2+} \times O^{2-}$	Ca_2O_2	$\Rightarrow CaO$
c.	Copper nitrate	$Cu^{2+} \times NO_3^{1-}$	$Cu_1(NO_3)_2$	$\Rightarrow Cu(NO_3)_2$
d.	Aluminium chloride	$Al^{3+} \times Cl^{1-}$	Al_1Cl_3	$\Rightarrow AlCl_3$
e.	Calcium carbonate	$Ca^{2+} \times CO_3^{2-}$	$Ca_2(CO_3)_2$	$\Rightarrow CaCO_3$

Self Assessment 2

1. What does an ionic compound contain? Give one example.
2. Write the chemical formula of a compound containing H^+ and SO_4^{2-} ions.
3. Identify the ions present in the compound $KMnO_4$. Name these ions.
4. Classify the following as acid radicals and basic radicals.

   (a) Al^{3+} (b) K^+ (c) SO_4^{2-} (d) NO_3^-
5. Show that covalency of H in H_2 is one.

REVIEW QUESTIONS

A. Very Short Answer Type Questions

1. How many valence electrons are there in an atom of oxygen?
2. Find the valency of metal M in the following compounds: (a) M_2O_3 and (b) MCl_3.
3. Write the two valencies shown by copper.
4. Which elements show positive valencies?
5. Name an element which shows a covalency of four.
6. Write the molecular formula of carbon dioxide.
7. Write the cationic and the anionic parts of the compound KCl.
8. Identify the ions present in the compound called ammonium carbonate.
9. What is the covalency of N atom in ammonia (NH_3)?
10. Write the chemical formula of a compound containing Na^+ and PO_4^{3-} ions.

B. Short Answer Type Questions

1. What are the valence electrons?
2. How are the molecular compounds formed?
3. What is the electrovalency of Na in NaCl? Explain.
4. Write the molecular formula of water. What is the ratio by mass of hydrogen and oxygen in water?
5. Classify the following as acid radicals and basic radicals:
   (a) NH_4^+ (b) HCO_3^- (c) Al^{3+} (d) CO_3^{2-} (e) SO_3^{2-}

C. Long Answer Type Questions

1. Define valency of an element in terms of:
   (a) its reaction with oxygen and chlorine
   (b) the electronic concept.
2. For three elements X, Y and Z, the following data are given.

Element	Mass number	No. of neutrons
X	35	18
Y	23	12
Z	24	12

   Give the chemical formulae and nature of compounds (ionic/covalent) formed between (a) X and X (b) X and Y (c) Z and X.
3. Describe the various steps involved in the writing of the chemical formula of an ionic compound called aluminium sulphate.
4. Write the chemical formulae of the compounds containing the following pairs of ions:
   (a) NH_4^+ and CO_3^{2-} (b) Zn^{2+} and PO_4^{3-} (c) Fe^{3+} and SO_4^{2-}
5. What is the variable valency? Name the following compounds in Stock notation:
   Cu_2O, CuO, FeO, Fe_2O_3 and SnO, SnO_2.
6. An element A has 4 valence electrons in its atom, whereas element B has only one valence electron in its atom. The compound formed by A and B does not conduct electricity. What is the nature of chemical bond in the compound formed? Give its electron dot structure.

D. True or False Type Questions

Write 'T' for True or 'F' for False.

1. Valency of an element represents combining capacity.
2. Electrovalency of chlorine in potassium chloride is 1+.
3. Valency of calcium in calcium carbonate is 2–.
4. Covalency of P in PH_3 is 3.
5. NH_4^+ is a basic radical.
6. Stock notation, Fe(III) represents ferric ion.
7. The molecular formula of hydrochloric acid is H_2Cl.
8. The carbonate ion is a divalent ion.
9. The sulphate ion is a polyatomic trivalent ion.
10. The chemical formula of ammonia is called its molecular formula.

E. Fill in the Blanks Type Questions

1. (a) Acid radicals possess ______________ charge.
   (b) Atoms of all elements except ______________ can lose or gain electrons.
   (c) Molecular compounds are formed by ______________ of electrons.
   (d) The name of K_2SO_3 is ______________ .
   (e) Basic radicals carry ______________ charge.
   (f) Chemical formula of zinc phosphate is ______________ .
   (g) Valency of oxygen is ______________ .
   (h) In Cu_2O, the valency of copper is ______________ .
   (i) The electrovalency of aluminium in aluminium sulphate is ______________ .
   (j) The covalency of nitrogen (N) in ammonia (NH_3) is ______________ .
2. Fill in the blanks in the following table.

	Compound	Cation	Anion	Molecular Formula
a.	Sodium hydroxide	Na^+	....	....
b.	....	Hg^{2+}	S^{2-}	....
c.	....	....	NO_3^-	$Al(NO_3)_2$
d.	....	Zn^{2+}	....	$ZnCO_3$
e.	Ferric sulphate	....	SO_4^{2-}	....
f.	Aluminium oxide	Al^{3+}	....	....
g.	....	....	HCO_3^-	$Mg(HCO_3)_2$
h.	Ammonium sulphite	....	SO_3^{2-}	....
i.	Potassium chromate	K^+	....	....
j.	....	Cr^{3+}	....	$CrPO_4$
k.	....	....	ClO_3^-	$KClO_3$
l.	Sodium thiosulphate	....	....	$Na_2S_2O_3$

3. Write chemical formula of each compound formed by the combination of corresponding cation and anion.

Basic radical		Chloride	Hydroxide	Nitrite	Sulphide	Sulphite
a.	Silver	....	....	....	....	....
b.	Sodium	....	....	....	....	....
c.	Potassium	....	....	....	....	....
d.	Ammonium	....	....	....	....	....
e.	Magnesium	....	....	....	....	....
f.	Calcium	....	....	....	....	....
g.	Ferrous	....	....	....	....	....
h.	Cadmium	....	....	....	....	....
i.	Ferric	....	....	....	....	....
j.	Aluminium	....	....	....	....	....

F. Match the Columns Type Questions

a.

Column A		Column B	
1	Au^{3+}	A.	Monovalent
2	Sulphite ion	B.	Anion
3	Chloride ion	C.	Trivalent
4	Mg^{2+}	D.	Tetravalent
5	Sn(IV)	E.	Cation

b.

Column A		Column B	
1	Ionic compound	A.	Carbon
2	Molecular formula	B.	+2, +3
3	Tetravalency	C.	Aluminium sulphate
4	Positive electrovalency	D.	CCl_4
5	Iron	E.	Sodium

G. Multiple Choice Type Questions

Choose the correct alternative.

1. Valency of iron (Fe) in Fe_2O_3 is
   (a) 2 ❑ (b) 3 ❑ (c) 1 ❑ (d) 5 ❑
2. Radicals carrying four units of charge are known as
   (a) divalent ❑ (b) trivalent ❑ (c) tetravalent ❑ (d) covalent ❑
3. $(NH_4)_2S$ is known as
   (a) ammonium sulphide ❑ (b) ammonium sulphite ❑
   (c) ammonium sulphate ❑ (d) ammonium bisulphate ❑
4. Molecular formula of hydrochloric acid is
   (a) ClH_2 ❑ (b) Cl_2H ❑ (c) HCl ❑ (d) HClO ❑
5. Correct chemical formula of zinc phosphate is
   (a) $Zn(PO_4)_2$ ❑ (b) $Zn_2(PO_4)_3$ ❑ (c) $Zn_3(PO_4)_2$ ❑ (d) Zn_2PO_4 ❑

6. Chemical name of Ag_2SO_4 is
   (a) silver sulphide ❑ (b) silver sulphite ❑
   (c) silver sulphate ❑ (d) silver thiosulphate ❑
7. Correct chemical formula of ammonium carbonate is
   (a) NH_4HCO_3 ❑ (b) $(NH_4)_2CO_3$ ❑ (c) $(NH_4)_3CO_2$ ❑ (d) $(NH_4)_2HCO_3$ ❑
8. Al_2O_3 is chemically known as
   (a) Dialuminium trioxide ❑ (b) Aluminium trioxide ❑
   (c) Aluminium oxide ❑ (d) Aluminium peroxide ❑
9. The covalency of carbon (C) in methane (CH_4) is
   (a) 1 ❑ (b) 2 ❑ (c) 3 ❑ (d) 4 ❑
10. Which of the following is not a polyatomic ion?
   (a) PO_4^{3-} ❑ (b) Cl^- ❑ (c) BO_3^{3-} ❑ (d) CO_3^{2-} ❑

ANSWERS

A. 1. Six 2. a. 3 b. 3 3. 1+, 2+ 4. Metals 5. Carbon 6. CO_2
7. K^+, Cl^- 8. NH_4^+, CO_3^{2-} 9. 3 10. Na_3PO_4

D. 1. T 2. F 3. F 4. T 5. T 6. T 7. F 8. T 9. F 10. T

E. 1. a. negative b. noble gases c. sharing d. potassium sulphite
e. positive f. $Zn_3(PO_4)_2$ g. 2 h. 1+
i. 3+ j. 3

2. a. OH^-, NaOH b. Mercury(II) sulphide, HgS
c. Aluminium nitrate, Al^{3+} d. Zinc carbonate, CO_3^{2-}
e. Fe^{3+}, $Fe_2(SO_4)_3$ f. O^{2-}, Al_2O_3
g. Magnesium hydrogen carbonate, Mg^{2+} h. NH_4^+, $(NH_4)_2SO_3$
i. CrO_4^{2-}, K_2CrO_4 j. Chromium phosphate, PO_4^{3-}
k. Potassium chlorate, K^+ l. Na^+, $S_2O_3^{2-}$

3.
(a)	AgCl,	AgOH,	$AgNO_2$,	Ag_2S,	Ag_2SO_3
(b)	NaCl,	NaOH,	$NaNO_2$,	Na_2S,	Na_2SO_3
(c)	KCl,	KOH,	KNO_2,	K_2S,	K_2SO_3
(d)	NH_4Cl,	NH_4OH,	NH_4NO_2,	$(NH_4)_2S$,	$(NH_4)_2SO_3$
(e)	$MgCl_2$,	$Mg(OH)_2$,	$Mg(NO_2)_2$,	MgS,	$MgSO_3$
(f)	$CaCl_2$,	$Ca(OH)_2$,	$Ca(NO_2)_2$,	CaS,	$CaSO_3$
(g)	$FeCl_2$,	$Fe(OH)_2$,	$Fe(NO_2)_2$,	FeS,	$FeSO_3$
(h)	$CdCl_2$,	$Cd(OH)_2$,	$Cd(NO_2)_2$,	CdS,	$CdSO_3$
(i)	$FeCl_3$,	$Fe(OH)_3$,	$Fe(NO_2)_3$,	Fe_2S_3,	$Fe_2(SO_3)_3$
(j)	$AlCl_3$,	$Al(OH)_3$,	$Al(NO_2)_3$,	Al_2S_3,	$Al_2(SO_3)_3$

F. a. 1 – C; 2 – B; 3 – A; 4 – E; 5 – D
b. 1 – C; 2 – D; 3 – A; 4 – E; 5 – B

G. 1. b 2. c 3. a 4. c 5. c 6. c 7. b 8. c 9. d 10. b

4

Acids and Bases

ACID – A hydrogen-containing compound which gives hydrogen ions (H_3O^+ or H^+) when dissolved in water, is called an **acid.** Acids can be classified as:

- Naturally-occurring acids
- Mineral acids

Mineral acids can be classified into two groups:

HYDRACIDS – The solutions of compounds of hydrogen with highly electronegative (nonmetallic) elements other than oxygen, are called **hydracids.**

OXOACIDS – The acids which contain an oxygen-containing anion (called **oxoanion**), are called oxoacids (earlier called **oxyacids**).

Thus, oxoacids are the compounds which contain hydrogen, oxygen and another nonmetallic element.

STRONG AND WEAK ACIDS

STRONG ACIDS – Acids which are completely ionised/dissociated when dissolved in water are called **strong acids.**

Example: Sulphuric acid (H_2SO_4), Hydrochloric acid (HCl) and Perchloric acid ($HClO_4$).

WEAK ACIDS – Acids which are partially ionised when dissolved in water are called **weak acids.**

Example: Ethanoic acid (CH_3COOH), Carbonic acid (H_2CO_3) and Hydrocyanic acid (HCN).

FORMULA OF A HYDRACID

The formula of a hydracid can be written as follows:

Step 1. Write the symbol of hydrogen (H).

Step 2. Write the symbol of the nonmetallic element present in the acid on the right of symbol H.

Step 3. Write the valency of hydrogen (+ 1) over H atom as **superscript.**

Step 4. Write the valency of the nonmetallic element over its symbol as the **superscript.**

Step 5. Criss-cross their valency numbers.

Step 6. Divide these numbers (written as subscripts) by a common factor, if required.

Step 7. Drop the subscript if '1', and write the formula of the hydracid.

Example: The formula of hydrochloric acid can be written as follows:

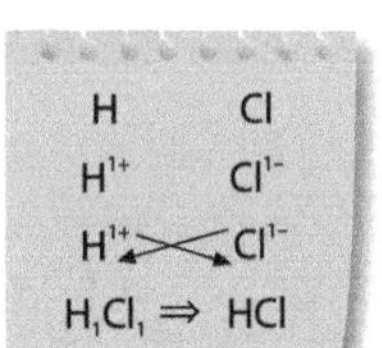

- Hydrochloric acid is made up of hydrogen and chlorine.
- Write the symbols of hydrogen and chlorine side-by-side.
- Write the valencies of hydrogen and chlorine as their superscripts.
- Criss-cross their valency numbers.
- Drop the subscripts (being 1 in both the cases).

  Thus, the formula of hydrochloric acid is HCl.

NAMING A HYDRACID FROM ITS FORMULA

Hydracids are the solutions of the corresponding hydracid gases in water. Such gases are called acid anhydrides (acid without water).

NAMING HYDRACID GASES

The hydracid gases are named as:

Name of the gaseous hydracid = Hydrogen + Name of the anion of the second element

Example: Name of HCl(g) = Hydrogen + Chloride → Hydrogen chloride (hydracid gas)

The gaseous HBr and H_2S are similarly named as hydrogen bromide and hydrogen sulphide respectively.

NAMING HYDRACIDS

Hydracids are named as follows:

Step 1. Write the word **Hydro** followed by the **root** for the name of second element.

Step 2. Add **-ic** and finally the term **acid** with a space (blank) between ic and acid.

Then,

Name of hydracid ⇒ Hydro + Root of the second element + ic + acid

Example:

- The name of HCl(*aq*) is obtained as follows:

$$\text{Name of HCl}(aq) = \text{HCl}(aq) \Rightarrow \underset{\text{Hydro}}{\text{Hydro}} + \text{chlor} + \text{ic} + \text{acid} \Rightarrow \text{Hydrochloric acid}$$

Hydro ↑ (from H); ↓ **chlor (root of the element chlorine)**

- The name of HCN(*aq*) is obtained as follows:

$$\text{Name of HCN}(aq) = \text{HCN} \Rightarrow \text{Hydro} + \text{cyan} + \text{ic} + \text{acid} \Rightarrow \text{Hydrocyanic acid}$$

Hydro ↑ (from H); ↓ **cyan (root of the CN group)**

- The name of HI(*aq*) is obtained as follows:

$$\text{Name of HI}(aq) = \text{HI} \Rightarrow \text{Hydr} + \text{iod} + \text{ic} + \text{acid} \Rightarrow \text{Hydriodic acid}$$

Hydr ↑ (from H); ↓ **iod (root of the element iodine)**

- HCl(*g*) is called **hydrogen chloride,** while HCl(*aq*) is called **hydrochloric acid.**
- HCN(*g*) is called **hydrogen cyanide,** while HCN(*aq*) is called **hydrocyanic acid.**
- HI(*g*) is called **hydrogen iodide,** while HI(*aq*) is called **hydriodic acid.**

The exception is HI(*aq*) in which case "o" in "hydro" is dropped while writing its name.

Name of some typical hydracids is given below:

Names of Some Typical Hydracids

Acid (gaseous form)		Acid (solution form)	
Formula	**Name**	**Formula**	**Name**
HF(*g*)	Hydrogen fluoride	HF(*aq*)	Hydrofluoric acid
HCl(*g*)	Hydrogen chloride	HCl(*aq*)	Hydrochloric acid
HBr(*g*)	Hydrogen bromide	HBr(*aq*)	Hydrobromic acid
HI(*g*)	Hydrogen iodide	HI(*aq*)	Hydriodic acid
$H_2S(g)$	Hydrogen sulphide	*$H_2S(aq)$	Hydrosulphuric acid
HCN(*g*)	Hydrogen cyanide	HCN(*aq*)	Hydrocyanic acid

*$H_2S(aq)$ is also named as sulphydric acid.

The distinction in naming the anhydrides and the acids is not critical for oxoacids, because all their anhydrides are different molecules. For example, the anhydride of H_2SO_4 is SO_3, not gaseous H_2SO_4. Thus, $H_2SO_4(aq)$ is always called sulphuric acid, not hydrogen sulphate.

Self Assessment 1

1. What are the compounds of hydrogen with any nonmetallic element called?
2. Name one hydracid.
3. Name the elements present in an oxoacid.
4. Name a gaseous covalent compound which when dissolved in water gives hydrochloric acid.
5. Name the root of the nonmetallic element present in HI acid.

WRITING THE FORMULA OF OXOACIDS

The formula of oxoacids can be written from the formula of the corresponding **oxoanion** and the sufficient number of hydrogen ions (H^+) so as to become a neutral molecule. Thus, we can write

Formula of oxoacid = $nH^+(aq)$ + Formula of oxoanion $\Rightarrow H_n$ oxoanion

Example: The formula of nitric acid can be written as follows:

- Nitric acid contains oxoanion called nitrate ion (NO_3^-).
- Nitrate ion (NO_3^-) carries one negative charge on it. So, it requires only one hydrogen ion (H^+) to form a neutral molecule.

Then,

Formula of the oxoacid containing NO_3^- ion = $H^+ + NO_3^- \Rightarrow HNO_3$

Formulae of Some Oxoacids

Oxoanion		Formula of the oxoacid						
Name	Formula							
Sulphate	SO_4^{2-}	$2H^+$	+	SO_4^{2-}	$\Rightarrow$	$H_2(SO_4)_1$	$\Rightarrow$	H_2SO_4
Carbonate	CO_3^{2-}	$2H^+$	+	CO_3^{2-}	$\Rightarrow$	$H_2(CO_3)_1$	$\Rightarrow$	H_2CO_3
Phosphate	PO_4^{3-}	$3H^+$	+	PO_4^{3-}	$\Rightarrow$	$H_3(PO_4)_1$	$\Rightarrow$	H_3PO_4

NAMING THE OXOACIDS FROM THEIR FORMULAE

Oxoacids contains one or more hydrogen atoms and the corresponding oxoanion.

- Each oxoanion is identified by its **word root** given in table (next page).

  To name any oxoacid proceed as follows:

Step 1. Identify and name the oxoanion in the acid molecule.

Step 2. Find the word root of the oxoanion by using its first part.

Step 3. Add the suffix.

Step 4. Name the oxoacid as follows:

Name of the oxoacid = Word root + Suffix + acid

Formulae of Some Oxoacids, the Names, Formulae and Roots of the Corresponding Oxoanions

Formula of oxoacid	Oxoanion		
	Name	Formula	Root
HNO_3	Nitrate	NO_3^-	Nitr
HNO_2	Nitrite	NO_2^-	Nitr
$HClO_4$	Perchlorate	ClO_4^-	Chlor
$HClO_3$	Chlorate	ClO_3^-	Chlor
$HClO_2$	Chlorite	ClO_2^-	Chlor
$HClO$	Hypochlorite	ClO^-	Chlor
H_2SO_4	Sulphate	SO_4^{2-}	Sulphur
H_2SO_3	Sulphite	SO_3^{2-}	Sulphur
H_3PO_4	Phosphate	PO_4^{3-}	Phosphor
H_2PO_3	Phosphite	PO_3^{2-}	Phosphor
H_3PO_2	Hypophosphite	PO_2^{3-}	Phosphor

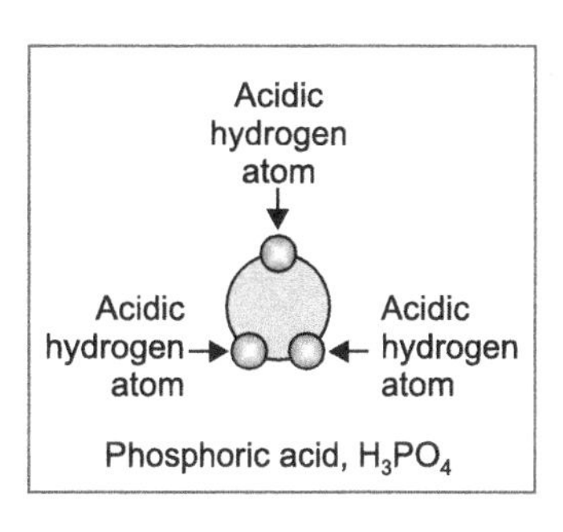

Phosphoric acid, H_3PO_4

Naming of some typical oxoacids is illustrated below:

- **For an oxoacid containing the most common oxoanion of its group use the suffix ic and add the term acid.**

  Examples:
  - Name of HNO_3 ⇒ Nitr + ic + acid ⇒ Nitric acid
  - Name of $HClO_3$ ⇒ Chlor + ic + acid ⇒ Chloric acid
  - Name of H_2SO_4 ⇒ Sulphur + ic + acid ⇒ Sulphuric acid
  - Name of H_3PO_4 ⇒ Phosphor + ic + acid ⇒ Phosphoric acid

- **For an acid containing the oxoanion ion with one less oxygen than the ic acid, use the suffix ous and add the term acid at the end.**

  Examples:
  - Name of HNO_2 ⇒ Nitr + ous + acid ⇒ Nitrous acid
  - Name of $HClO_2$ ⇒ Chlor + ous + acid ⇒ Chlorous acid
  - Name of H_2SO_3 ⇒ Sulphur + ous + acid ⇒ Sulphurous acid
  - Name of H_3PO_3 ⇒ Phosphor + ous + acid ⇒ Phosphorous acid

- **For an acid containing the polyatomic oxoanion with two less oxygens than the ic acid, use the prefix hypo and the suffix ous, and add the term acid at the end.**

  Examples:

  - Name of $HClO$ ⇒ Hypo + chlor + ous + acid ⇒ Hypochlorous acid
  - Name of H_3PO_2 ⇒ Hypo + phosphor + ous + acid ⇒ Hypophosphorous acid

- **For an acid containing the polyatomic ion with one more oxygen than the ic acid, use the prefix per and the suffix ic, and add the term acid at the end.**

  Examples:

  - Name of $HClO_4$ ⇒ Per + chlor + ic + acid ⇒ Perchloric acid
  - Name of H_2SO_5 ⇒ Per + sulphur + ic + acid ⇒ Persulphuric acid

HSO_3^- is not an acid molecule; it is an anion because it carries a 1– charge. Even though, it shows acidic properties, it is named like a polyatomic anion. For a H-containing chemical species to be called an acid, its molecule must not contain metal atoms.

$NaHSO_3$ should not be named as an acid. Instead, it should be named as a compound because it consists of a Na^+ cation and an HSO_3^- anion. Thus, it is named sodium bisulphite or sodium hydrogensulphite.

Some Exceptions

Notice that the whole name for sulphur, and not just the root, sulph-, is found in the name sulphuric acid.

Similarly, although the usual root for phosphorus is **phosph–**, the root **phosphor–** is used for phosphorus-containing oxoacids, as in the name phosphoric acid.

Self Assessment 2

1. Write the general formula of an oxoacid.
2. Identify the oxoanion in sulphuric acid molecule.
3. What is the root for phosphate oxoanion?
4. Can $NaHSO_3$ be called an acid? Give reason.
5. Write the formula of hypochlorous acid.

BASES AND ALKALIS

BASE – A **base** is a compound which gives free hydroxide ions (OH^-) when dissolved in water.

Example: Some common bases are:

- Sodium hydroxide ($NaOH$)
- Ammonia (NH_3)
- Sodium carbonate (Na_2CO_3)
- Copper hydroxide [$Cu(OH)_2$]
- Lime (CaO)
- Urea ($NH_2.CO.NH_2$)

Bases can be classified as **strong bases** and **weak bases.**

Example: Strong bases : NaOH, KOH, $Ba(OH)_2$

Weak bases : NH_3, $NH_2.CO.NH_2$, $Cu(OH)_2$, CaO, $Ca(OH)_2$

ALKALI – The strong bases which are highly soluble in water and contain one or more hydroxide ions (OH^-) in their molecules are called **alkalis.**

Example: Some common alkalis are:

- Sodium hydroxide (NaOH)
- Potassium hydroxide (KOH)
- Barium hydroxide ($Ba(OH)_2$)
- Calcium hydroxide ($Ca(OH)_2$)

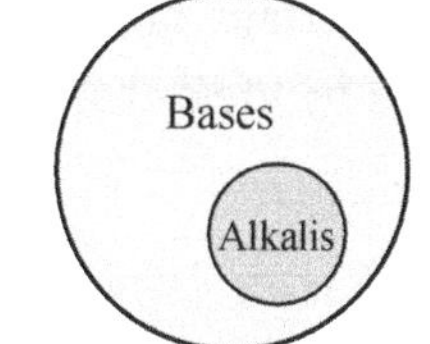

Venn diagram showing the relationship between bases and alkalis

NAMING BASES AND ALKALIS

Bases and alkalis are named on the basis of the type of their molecular formulae.

Naming Hydroxy Bases

The bases containing one or more hydroxide ions (OH^-) in their molecules are called **hydroxy bases.** These are named as follows:

> Remember!
> All alkalis are Bases whereas all bases are not Alkalis.

- Identify **positive ion** in the molecule of the base.
- Write the name of the positive ion followed by the word **hydroxide.**

Thus,

Name of a hydroxy base ⇒ Name of the positive ion + hydroxide

Example: NaOH is named as follows:

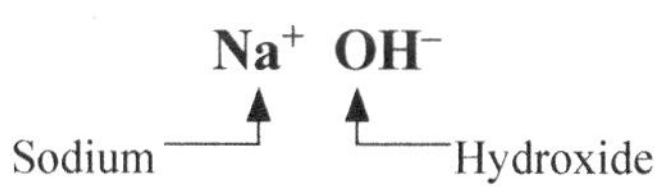

Name of NaOH ⇒ Sodium + hydroxide ⇒ Sodium hydroxide

Similarly, we can write the name of some more hydroxy bases. These are:

Name of KOH ⇒ Potassium + hydroxide ⇒ Potassium hydroxide

Name of $Ca(OH)_2$ ⇒ Calcium + hydroxide ⇒ Calcium hydroxide

Name of NH_4OH ⇒ Ammonium + hydroxide ⇒ Ammonium hydroxide

Naming Oxide Bases

Most simple oxides of metals are **basic** in nature. These are named as follows:

Example: CaO is named as follows:

Ca^{2+} O^{2-}

Calcium Oxide

Name of an oxide base ⇒ Name of the metal present in the base + oxide

- Name of CaO ⇒ Calcium + oxide ⇒ Calcium oxide
- Name of MgO ⇒ Magnesium + oxide ⇒ Magnesium oxide
- Name of Na_2O ⇒ Sodium + oxide ⇒ Sodium oxide

Naming Nitrogenous Bases

Certain nitrogen-containing molecular compounds are basic in nature. These are called by their characteristic names.

Example:

- NH_3 is named as Ammonia.
- $NH_2 \cdot CO \cdot NH_2$ is named as Urea.
- $C_6H_5NH_2$ is named as Aniline.

Self Assessment 3

1. Name a weak base.
2. Name an alkali.
3. Write the formula of ammonium hydroxide.
4. Write the formula of magnesium oxide.
5. Is $Cu(OH)_2$ a base or an alkali?

REVIEW QUESTIONS

A. Very Short Answer Type Questions

1. Give an example of a hydracid.
2. Name an oxoacid in which the anion carries 2– charge.
3. Name two strong acids.
4. What is the root for chlorine as the second element in its hydracid?
5. Write formula of the oxoacid containing nitrate ion.
6. Write name of the oxoacid H_3PO_4.
7. What is the root for sulphate ion in sulphuric acid?
8. Write the formula of hypochlorous acid.
9. Give one example of a base which cannot be classified as an alkali.
10. Give one example of a nitrogenous base.

B. Short Answer Type Questions

1. Differentiate between chloric and chlorous acids.
2. Why do solutions of acids conduct electricity?
3. Give valency of the highlighted (in bold letter) symbol of the element in the following molecules.

   (a) NaOH (b) H_2CO_3 (c) H_3PO_4.
4. What are oxoacids?
5. What are the acid anhydrides? Name the acid anhydride of sulphuric acid.

C. Long Answer Type Questions

1. Write the formulae of the following acids:
   (a) Carbonic acid
   (b) Hydroselenic acid
   (c) Hydrofluoric acid
   (d) Hydrosulphuric acid
   (e) Sulphurous acid
   (f) Hydrocyanic acid.
2. How are oxoacids named on the basis of their formulae? Give one example.
3. How are the hydroxy bases named? Give one example.
4. What are the nitrogenous bases? Name three such bases.
5. Name the following bases:

   (a) NaOH (b) $Ca(OH)_2$ (c) NH_3

   (d) $Cu(OH)_2$ (e) BaO (f) $Be(OH)_2$.

D. True or False Type Questions

Write 'T' for True or 'F' for False.

1. Ammonia is an acid.
2. Hydracids contain only hydrogen.
3. The solutions of acids in water are good conductors of electricity.
4. A base is a compound which gives free H^+ ions when dissolved in water.
5. MgO is basic in nature.
6. Solution of HCl gas in water is called hydrochloric acid.
7. $NaHSO_3$ is an acid.
8. Oxoanion in sulphurous acid is SO_3^{2-}.
9. Strong acids are also called alkalis.
10. Urea is an example of a nitrogenous base.

E. Fill in the Blanks Type Questions

1. Complete the following table. (One set has been solved)

Oxoacid		Oxoanion			
Formula	Name	Formula	Charge	Valence	Name
$HClO_3(aq)$	Chloric acid	ClO_3^-	1–	1	Chlorate
$HClO_4(aq)$	Perchloric acid	ClO_4^-	1–	1	Perchlorate
$HClO_2(aq)$	Chlorous acid	ClO_2^-	1–	1	Chlorite
$HClO(aq)$	Hypochlorous acid	ClO^-	1–	1	Hypochlorite
$H_2SO_5(aq)$	Persulphuric acid	SO_5^{2-}	2–	2	Persulphate
$HBrO_3(aq)$	....	....	....	....	....
$HBrO_2(aq)$	....	....	....	....	....
$HIO(aq)$	....	....	....	....	....
$HIO_4(aq)$	....	....	....	....	....
$H_2CO_3(aq)$	....	....	....	....	....
$H_2SO_4(aq)$	....	....	....	....	....
$H_2SO_3(aq)$	....	....	....	....	....
$H_3PO_2(aq)$	....	....	....	....	....
$H_3PO_3(aq)$	....	....	....	....	....
$H_3PO_4(aq)$	....	....	....	....	....
$H_3PO_5(aq)$	....	....	....	....	....

2. Complete the following table.

Acid formula	Second element	Root for the second element	Name of the acid
$HCl(aq)$	....	....	....
$HF(aq)$	....	....	....
$H_2S(aq)$	....	....	....
$HBr(aq)$	....	....	....
$HI(aq)$	....	....	....

3. Fill in the names of the bases in the following table.

Base formula	Name of the base
$NaOH$	....
Li_2O	....
KOH	....
CaO	....
$Mg(OH)_2$	....

F. Match the Columns Type Questions

a.

	Column A		Column B
1	Hydrosulphuric acid	A.	HF(*aq*)
2	Acetic acid	B.	$H_2S(aq)$
3	Hypobromous acid	C.	HBrO
4	Hydrofluoric acid	D.	H_3PO_4
5	Phosphoric acid	E.	CH_3COOH

b.

	Column A		Column B
1	Perbromic acid	A.	H_2CO_3
2	Iodous acid	B.	$HClO_3$
3	Chloric acid	C.	HIO_2
4	Carbonic acid	D.	HI(*aq*)
5	Hydriodic acid	E.	$HBrO_4$

c.

	Column A		Column B
1	Hydrobromic acid	A.	H_2SO_5
2	Iodic acid	B.	HNO_2
3	Chlorous acid	C.	HBr(*aq*)
4	Nitrous acid	D.	HIO_3
5	Persulphuric acid	E.	$HClO_2$

G. Multiple Choice Type Questions

Choose the correct alternative.

1. The correct formula of hyposulphurous acid is
   (a) H_2SO ❑ (b) H_2SO_2 ❑ (c) H_2SO_3 ❑ (d) H_2SO_4 ❑
2. The correct formula of the base calcium hydroxide is
   (a) KaOH ❑ (b) KOH ❑ (c) $Ca(OH)_2$ ❑ (d) $C(OH)_4$ ❑
3. Chloric acid can be written as
   (a) ClOH ❑ (b) $HClO_2$ ❑ (c) $HClO_3$ ❑ (d) HCl ❑
4. The correct formula of persulphuric acid is
   (a) H_2SO_5 ❑ (b) H_2SO_4 ❑ (c) H_2SO_3 ❑ (d) H_2SO_2 ❑
5. Which of the following acids contain anion with a valency 1– ?
   (a) H_2CO_3 ❑ (b) H_2SO_5 ❑ (c) H_3PO_2 ❑ (d) HNO_2 ❑
6. Which of the following is not a nitrogenous base?
   (a) NH_3 ❑ (b) HNO_2 ❑ (c) $C_6H_5NH_2$ ❑ (d) Urea ❑
7. Charge on the ammonium ion in ammonium phosphate is
   (a) 1– ❑ (b) 1+ ❑ (c) 3+ ❑ (d) 3– ❑

8. Which of the following is not a diprotic acid?
   (a) Carbonic acid ❑ (b) Sulphuric acid ❑
   (c) Hyposulphuric acid ❑ (d) Chlorous acid ❑
9. An example of a weak base is
   (a) NaOH ❑ (b) $Sr(OH)_2$ ❑
   (c) $Ba(OH)_2$ ❑ (d) $Cu(OH)_2$ ❑
10. Which of the following is an oxide base?
   (a) $Ca(OH)_2$ ❑ (b) MgO ❑
   (c) $Cu(OH)_2$ ❑ (d) NH_4OH ❑

ANSWERS

A. 1. Hydrochloric acid 2. Sulphuric acid 3. Sulphuric acid, Hydrochloric acid 4. Chlor
5. HNO_3 6. Phosphoric acid 7. Sulphur
8. HClO 9. $Cu(OH)_2$ 10. Ammonia (NH_3)

D. 1. F 2. F 3. T 4. F 5. T 6. T 7. F 8. T 9. F 10. T

E. 1.

Bromic acid,	BrO_3^-,	1–,	1,	Bromate
Bromous acid,	BrO_2^-,	1–,	1,	Bromite
Hypoiodous acid,	IO^-,	1–,	1,	Hypoiodite
Periodic acid,	IO_4^-,	1–,	1,	Periodate
Carbonic acid,	CO_3^{2-},	2–,	2,	Carbonate
Sulphuric acid,	SO_4^{2-},	2–,	2,	Sulphate
Sulphurous acid	SO_3^{2-},	2–,	2,	Sulphite
Hypophosphorous acid,	PO_2^{3-},	3–,	3,	Hypophosphite
Phosphorous acid,	PO_3^{3-},	3–,	3,	Phosphite
Phosphoric acid,	PO_4^{3-},	3–,	3,	Phosphate
Peroxomonophosphoric acid,	PO_5^{3-},	3–,	3,	Peroxomonophosphate

2.

Chlorine,	Chlor,	Hydrochloric acid
Fluorine,	Fluor,	Hydrofluoric acid
Sulphur,	Sulphur,	Hydrosulphuric acid
Bromine,	Brom,	Hydrobromic acid
Iodine,	Iod,	Hydriodic acid

3. Sodium hydroxide, Lithium oxide, Potassium hydroxide, Calcium oxide, Magnesium hydroxide.

F. a. 1 – B; 2 – E; 3 – C; 4 – A; 5 – D
b. 1 – E; 2 – C; 3 – B; 4 – A; 5 – D
c. 1 – C; 2 – D; 3 – E; 4 – B; 5 – A

G. 1. b 2. c 3. c 4. a 5. d 6. b 7. b 8. d 9. d 10. b

5

Salts

SALTS – An ionic compound which when dissolved in water, breaks up completely into its constituent ions is called a salt.

- A salt is always formed when an acid is neutralised by a base.

  Acid + Base → Salt + Water

- The name of a salt contains two parts:
  - The first part (the **cation)** comes from the parent base.
  - The second part (the **anion)** comes from the parent acid.

Example:

- The two parts of **Potassium nitrate are:**
  - **Potassium ion** – Coming from the parent base potassium hydroxide.
  - **Nitrate ion** – Coming from the parent acid nitric acid.
- The two parts of **Copper(II) sulphate are:**
  - **Copper ion** – Coming from the parent base called copper hydroxide.
  - **Sulphate ion** – Coming from the parent acid called sulphuric acid.

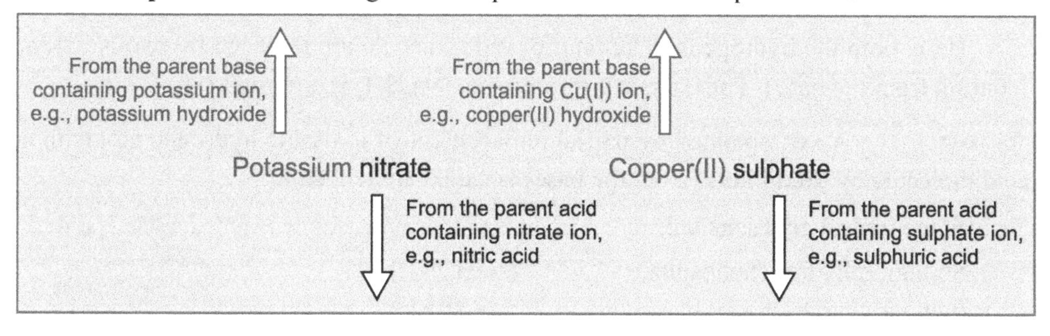

TYPES OF SALTS

Salts are broadly classified into the following three categories:

- Normal salt
- Acid salt
- Basic salt

These salts are described below:

NORMAL SALT – A salt obtained by complete replacement of the ionisable hydrogen atoms in an acid molecule by **metal ions,** or **ammonium ion** is called a **normal salt.**

Some typical normal salts are:

- Sodium chloride ($NaCl$)
- Sodium sulphate (Na_2SO_4)
- Calcium carbonate ($CaCO_3$)
- Sodium carbonate (Na_2CO_3)

Formation of Sodium Chloride

Sodium chloride is a salt formed by the reaction of hydrochloric acid ($HCl(aq)$) with sodium hydroxide ($NaOH$).

$$HCl(aq) + NaOH(aq) \rightarrow NaCl(aq) + H_2O(l)$$

hydrochloric acid (acid) — sodium hydroxide (alkali) — sodium chloride (normal salt) — water

Here, the acid (HCl) molecule contains only one ionisable H-atom. This ionisable H-atom is replaced by one Na^+ ion coming from the base $NaOH$ to form the salt $NaCl$. Therefore, the salt $NaCl$ (sodium chloride) is a normal salt.

Formation of Sodium Sulphate

Sodium sulphate is obtained from sulphuric acid (H_2SO_4) and sodium hydroxide ($NaOH$).

$$2NaOH + H_2SO_4 \rightarrow Na_2SO_4 + 2H_2O$$

2H atoms in H_2SO_4 are replaced by 2Na from NaOH — sodium sulphate (normal salt)

Here, both the hydrogens in sulphuric acid molecule are replaced by two Na atoms (coming from the base). Thus, the salt formed, i.e., Na_2SO_4 is a normal salt.

ACID SALT – A salt obtained by partial replacement of ionisable hydrogen atoms in an acid molecule by **metal ions** (from the base) is called an **acid salt.**

An acid salt is formed when one or more hydrogens in the acid molecule are left unreacted.

Some typical acid salts are:

- Sodium dihydrogenphosphate (NaH_2PO_4)
- Sodium monohydrogenphosphate (Na_2HPO_4)
- Sodium monohydrogensulphate ($NaHSO_4$)

Formation of Sodium Hydrogencarbonate

Sodium hydrogencarbonate (an acid salt) is formed when only one ionisable H atom of carbonic acid (H_2CO_3) is replaced by one sodium ion (coming from the base).

$H_2CO_3\,(aq)$	$-$	H^+	$+$	Na^+	$\rightarrow$	$NaHCO_3(aq)$
carbonic acid (contains two H atoms)				from the base NaOH		sodium hydrogencarbonate (or sodium bicarbonate) ($NaHCO_3$ is an **acid salt)**

BASIC SALT – A salt which contains one or more hydroxy groups (coming from a base) alongwith the anion coming from the parent acid is called a **basic salt.**

Some typical basic salts are:

- Lead hydroxychloride [Pb(OH)Cl] (Basic lead chloride)
- Basic copper carbonate [$CuCO_3.Cu(OH)_2$] (Malachite)
- Calcium hydroxychloride [Ca(OH)Cl]

Formation of Basic Lead Chloride

A basic salt is formed when one or more unreacted -OH^- remain in the base molecule.

$Pb(OH)_2$	$-$	OH^-	$+$	Cl^-	$\rightarrow$	$Pb(OH)Cl$
lead hydroxide (contains two OH^- ions)				from HCl		basic lead chloride (a basic salt) (contains one OH^- ion)

NAMING SALTS FROM THEIR FORMULAE

The name of a salt from its formula is written as follows:

Step 1. Split the formula of the salt into two parts — one coming from the **parent base** and other from the **parent acid.**

Step 2. The part coming from the parent base is the **cation** (positive ion) and that coming from the parent acid is the **anion** (negative ion).

Step 3. Write the names of these ions — the cation first, then anion, i.e.,

Name of the salt $\Rightarrow$ **Name of the cation from the base** + **Name of the anion from the acid**

NAMING NORMAL SALTS

Naming of some normal salts is illustrated below:

- NaCl $\Rightarrow$ Na^+ (coming from the base) + Cl^- (coming from the acid) $\Rightarrow$ Sodium + Chloride $\Rightarrow$ Sodium chloride
- KNO_3 $\Rightarrow$ K^+ (coming from the base) + NO_3^- (coming from the acid) $\Rightarrow$ Potassium + Nitrate $\Rightarrow$ Potassium nitrate
- Na_2CO_3 $\Rightarrow$ $2Na^+$ (coming from the base) + CO_3^{2-} (coming from the acid) $\Rightarrow$ Sodium + Carbonate $\Rightarrow$ Sodium carbonate

NAMING ACID SALTS

Naming of some acid salts is illustrated below:

- $NaHSO_4$ (acid salt) ⇒ Na^+ (coming from the base NaOH) + HSO_4^- (coming from the acid H_2SO_4) ⇒ Sodium + Hydrogensulphate ⇒ Sodium hydrogensulphate
- $NaHCO_3$ ⇒ Na^+ (coming from the base NaOH) + HCO_3^- (coming from the acid H_2CO_3) ⇒ Sodium + Hydrogencarbonate ⇒ Sodium hydrogencarbonate
- NaH_2PO_4 ⇒ Na^+ (coming from the base NaOH) + $H_2PO_4^-$ (coming from the acid H_3PO_4) ⇒ Sodium + Dihydrogenphosphate ⇒ Sodium dihydrogenphosphate

> While naming an acid salt, an appropriate **prefix** should be added before the word 'hydrogen' depending upon the number of hydrogen atoms present in the salt, viz.
>
> - **mono** for **one**
> - **di** for **two**
> - **tri** for **three**

NAMING BASIC SALTS

Basic salts are named by adding the word **'hydroxy'** between the **cation** (coming from the parent base) and the **anion** (coming from the parent acid).

The basic salt Pb(OH)Cl is named as:

Lead + hydroxy + chloride ⇒ Lead hydroxychloride

(Lead: From the parent base; chloride: From the parent acid)

Self Assessment

1. Name an acid which gives both normal as well as acid salts.
2. Name the following normal salts:
   (a) CaO (b) FeS (c) NaH (d) $BaSO_4$.
3. What is the nature of an aqueous solution of an acid salt?
4. Name the following salts:
   (a) Na_2HPO_4 (b) Pb(OH)Cl (c) Ca(OH)Cl (d) NaH_2PO_4.
5. Name the parent acid of the salt Na_3PO_4.

REVIEW QUESTIONS

A. Very Short Answer Type Questions

1. Give one example of a normal salt.
2. What type of salt is Na_2HPO?
3. Give the correct name of $Fe_2(HPO_4)_3$.

4. What is the charge on the anion in lithium carbonate?
5. Name the cation and anion present in the molecule of magnesium hydrogenphosphate.
6. Name the salt having molecular formula $KMnO_4$.
7. What type of salt is the compound magnesium perchlorate?
8. Is barium sulphate a basic salt?
9. Is NH_4OH an acid or a base?
10. Write the formula of parent acid of the salt NaOCN.

B. Short Answer Type Questions

1. Identify the parent bases of the following salts and name them:
   (a) $Al_2(SO_4)_3$ (b) $CuSO_4$ (c) $LiNO_3$ (d) $BaSO_4$.
2. Identify the parent acids of the following salts and name them:
   (a) Na_2CO_3 (b) $CaCO_3$ (c) KBr (d) NH_4Cl.
3. Give the names of the following salts:
   (a) Cu_2S (b) FeI_2 (c) KBr (d) NH_4Cl
   (e) $Fe_2(SO_4)_3$ (f) $AlPO_4$ (g) CaF_2 (h) $AgNO_3$.
4. Write the correct chemical formulae of the following salts:
   (a) Calcium nitrate (b) Aluminium sulphate (c) Sodium cyanide
   (d) Potassium chloride (e) Lithium bromide (f) Copper(II) sulphide
   (g) Basic zinc carbonate.
5. Show the steps involved in naming the salt sodium iodate, identifying its parent acid and the parent base.

C. Long Answer Type Questions

1. What are salts? How are salts formed?
2. All salts contain two parts. Identify two parts of the salt potassium chromate.
3. How does an acid salt differ from a normal salt? Write the formulae of the normal and the acid salts obtained from sulphuric acid and sodium hydroxide.
4. How are the salts named? Illustrate by giving an example.
5. What is a basic salt? How is a basic salt named?

D. True or False Type Questions

Write 'T' for True or 'F' for False.

1. All salts are ionic in nature.
2. The classification of salts is based on the pH value of their solutions.
3. Most sodium salts are soluble in water.
4. Salts when dissolved in water, dissociate to give free electrons.
5. Potassium cyanide can be formed by mixing calcium cyanide with potassium hydroxide.
6. A salt always contains anion coming from its parent base.
7. $KHCO_3$ is a basic salt.
8. Li_2CO_3 contains a monovalent cation and a divalent anion.
9. In a normal salt, both the ions carry equal charge.
10. Normal salts when dissolved in water, always give neutral solutions.

E. Fill in the Blanks Type Questions

1. Write names of the given salts in blank space.

Formula of salt	Name of salt	Formula of salt	Name of salt
$Al(NO_2)_3$		$Ca(NO_3)_2$	
$AgIO_3$		KNO_3	
$AgNO_3$		KCN	
Ag_2SO_4		Li_2SO_3	
$Al(OH)_3$		$LiCl$	
$AlCl_3$		$MgSO_4$	
$Ba_3(PO_4)_2$		$Mg(OH)_2$	
$BaSO_3$		Na_3PO_3	
$CaCO_3$		Na_2SO_4	
CaF_2		$NaHCO_3$	
$NaCl$		$Ca_3(PO_4)_2$	
NH_4NO_3		CH_3COONa	

2.

Name of salt	Formula	Name of salt	Formula
Aluminium fluoride		Lithium nitrate	
Aluminium hydroxide		Lithium chloride	
Ammonium nitrite		Magnesium nitride	
Ammonium sulphite		Magnesium hydrogenphosphate	
Ammonium phosphite		Magnesium hydroxide	
Barium phosphide		Potassium cyanide	
Barium bromite		Potassium sulphate	
Calcium fluoride		Sodium acetate	
Calcium phosphate		Magnesium perchlorate	
Calciumhydrogencarbonate		Sodium sulphite	
Calcium sulphate		Sodium nitrate	

F. Match the Columns Type Questions

a.

	Column A		Column B
1	$Ca_3(PO_4)_2$	A.	H_2SO_4
2	$Zn(OH)_2$	B.	HCl
3	$CaSO_4$	C.	CaO
4	$CaCO_3$	D.	H_3PO_4
5	$MgCl_2$	E.	ZnO

b.

	Column A		Column B
1	Cation	A.	NaCl
2	Normal salt	B.	H_3PO_4
3	$NaHCO_3$	C.	HNO_2
4	$AlPO_4$	D.	Parent base
5	Ammonium nitrite	E.	Acid salt

G. Multiple Choice Type Questions

Choose the correct alternative.

1. The salt NaCl is a/an

   (a) normal salt ❑ (b) acid salt ❑ (c) basic salt ❑ (d) neutral salt ❑

2. The charge on cation of the salt $Ca_3(PO_4)_2$ is

   (a) 2+ ❑ (b) 3+ ❑ (c) 6 + ❑ (d) 0 ❑

3. A salt always contains a metal cation coming from the parent

   (a) acid ❑ (b) base ❑ (c) another salt ❑ (d) none of these ❑

4. The correct formula of copper(II) sulphide is

   (a) Cu_3S_2 ❑ (b) Cu_2S ❑ (c) CuS ❑ (d) CuS_2 ❑

5. Which of the following is/are acid salts?

   (a) NaCl ❑ (b) $NaHCO_3$ ❑ (c) Na_2CO_3 ❑ (d) $Fe_3(HPO_4)_2$ ❑

6. The correct formula of iron(III) hydrogenphosphate is

   (a) $Fe_4(HPO_4)_3$ ❑ (b) $Fe_2(HPO_4)_3$ ❑

   (c) $Fe_2(HPO_4)$ ❑ (d) $Fe_3(HPO_4)_2$ ❑

7. What type of salt iron(III) hydrogenphosphate is?

   (a) normal ❑ (b) acid ❑ (c) basic ❑ (d) none of these ❑

8. Which of the following has the anion with 3– charge?

   (a) $ZnCO_3$ ❑ (b) $AlPO_4$ ❑ (c) $CaSO_4$ ❑ (d) Na_2CO_3 ❑

9. The parent acid from which anion of the molecule sodium hydrogensulphite comes is

   (a) sulphuric acid ❑ (b) sulphurous acid ❑

   (c) persulphuric acid ❑ (d) hyposulphuric acid ❑

10. What type of salt is barium thiosulphate?

    (a) acid salt ❑ (b) basic salt ❑ (c) normal salt ❑ (d) complex salt ❑

ANSWERS

A. 1. NaCl　2. Acid salt　3. Fe(III) monohydrogenphosphate
4. 2–　5. Mg^{2+}, HPO_4^{2-}　6. Potassium permanganate
7. Normal　8. No　9. Base　10. HCNO

D. 1. F　2. F　3. T　4. F　5. T　6. F　7. F　8. T　9. T　10. F

E. 1. **Column-wise**

- Aluminium nitrite,　Silver iodate,　Silver nitrate,
  Silver sulphate,　Aluminium hydroxide,　Aluminium chloride,
  Barium phosphate,　Barium sulphite,　Calcium carbonate,
  Calcium fluoride,　Sodium chloride,　Ammonium nitrate.
- Calcium nitrate,　Potassium nitrate,　Potassium cyanide,
  Lithium sulphite,　Lithium chloride,　Magnesium sulphate,
  Magnesium hydroxide,　Sodium phosphate,　Sodium sulphate,
  Sodium hydrogencarbonate,　Calcium phosphate,　Sodium acetate.

2. **Column-wise**

- AlF_3,　$Al(OH)_3$,　NH_4NO_2,　$(NH_4)_2SO_3$,　$(NH_4)_2PO_3$,　Ba_3P_2,
  $Ba(BrO_2)_2$,　CaF_2,　$Ca_3(PO_4)_2$,　$Ca(HCO_3)_2$,　$CaSO_4$
- $LiNO_3$,　LiCl,　Mg_3N_2,　$Mg(HPO_3)_2$,　$Mg(OH)_2$,　KCN,
  K_2SO_4,　CH_3COONa,　$Mg(ClO_4)_2$,　Na_2SO_3,　$NaNO_3$

F. a. 1 – D;　2 – E;　3 – A;　4 – C;　5 – B
b. 1 – D;　2 – A;　3 – E;　4 – B;　5 – C

G. 1. a　2. a　3. b　4. c　5. b, d　6. b　7. b　8. b　9. b　10. c

Fun To Do

Ionic Sudoku

In Ionic Sudoku, use logic to work out the compounds in the blanks squares. Every 2 × 2 box, row and column contains a fluoride and compounds where the anions are

J^{a-}, Q^{b-} and Z^{c-}.

Al_2J_3	AlZ_3		K_3Q
	Ca_3Q_2		MgF_2
	CaJ		

Each **2 × 2 box** is based on compounds containing only **one cation.**

[**Hint:** Start by working out charges on all the ions.]

6

Chemical Equations

Chemical Reactions – The change of one or more substances into other substances having different composition and properties is called a **chemical reaction.**

- Substances which take part in a chemical reaction are called **reactants.**
- Substances which are formed in a chemical reaction are called **products.**

Exothermic Reactions – The term **exothermic** means giving out heat (*exo* – out and *thermic* – heat). So, a reaction in which heat is liberated (given out) is known as an **exothermic reaction**.

An exothermic reaction can be described as follows:

Reactants $\rightarrow$ Products + Heat

Some typical exothermic reactions are:

- Burning of coal/carbon
- Dissolution of concentrated sulphuric acid in water
- Dissolution of caustic soda (sodium hydroxide, NaOH) in water
- Combustion of LPG, CNG, Methane, Kerosene, Furnace oil, etc.
- Reaction of quicklime (CaO) with water (H_2O)

Endothermic Reactions – The term **endothermic** means absorbing (taking in) heat (*endo* – in and *thermic* – heat). So, a reaction in which heat is absorbed is known as an **endothermic reaction.**

An endothermic reaction can be described as follows:

Reactants + Heat $\rightarrow$ Products

or Reactants $\rightarrow$ Products – Heat

Some typical endothermic reactions are:

- Dissolution of ammonium chloride (NH_4Cl) in water
- Conversion of water to steam
- Reaction between nitrogen and oxygen to form nitric oxide

DESCRIBING A CHEMICAL REACTION

Chemical reactions can be described in the following two ways:

- In words, using the names of the reactants and products
- Using symbols and formulae of the reactants and products

WORD EQUATION – An equation expressed by describing the names of the reactants and products in words is called a **word equation.**

Example: The decomposition reaction of mercury(II) oxide into mercury and oxygen can be described by the word equation given below:

$$\text{Mercury(II) oxide} \xrightarrow{\text{heat}} \text{Mercury} + \text{Oxygen}$$

CHEMICAL EQUATION – A reaction expressed using symbols and formulae of the reactants and products is called a **chemical equation.**

or

A shorthand representation of a chemical reaction in terms of symbols and formulae of the substances involved is called a **chemical equation.**

Example: The above decomposition reaction can be described by a chemical equation given below:

$$2HgO(s) \xrightarrow{\text{heat}} 2Hg(l) + O_2(g)$$

WRITING A CHEMICAL EQUATION

The chemical equation for a chemical reaction is written as follows:

Step 1. Identify the reactants and products of the chemical reaction.

Step 2. Write down the formulae or symbols of the reactants on the left-hand side with a sign of plus (+) between them.

The formulae or symbols of the products formed in the reaction are written on the right-hand side with a sign of plus (+) between them.

The reactants and products are separated by an arrow (→) or by a sign of equality (=).

The chemical equation so obtained is called a **skeleton equation.**

Step 3. Count the number of atoms of each element on both the sides.

- If the number of atoms of each element on both the sides of the equation is **equal**, then it is called a **balanced chemical equation.**
- If the number of atoms of any one or more of the elements on both the sides is **not equal**, then the chemical equation is called an **unbalanced chemical equation.**
- The number of atoms of each element is made equal by adjusting the coefficients before the symbols and the formulae of the reactants and products.

Step 4. The process by which the number of atoms of each element on both the sides are made equal, is called the **balancing of chemical equation.**

Step 5. At the end, the chemical equation is made molecular, if required.

MAKING A CHEMICAL EQUATION MORE INFORMATIVE

A chemical equation can be made more informative by supplying additional information to the chemical equation. This is done as follows:

- Reaction conditions, e.g., **temperature** (t), **pressure** (p) and **catalyst** are shown over the arrow head. $\xrightarrow{t,\ p,\ \text{catalyst}}$
- Physical states of the reactants and products are described by the letters (s), (l), (g) or (aq) for solid, liquid, gas or aqueous solution respectively at the end of the formula/symbol. These letters are known as **state symbols.**

  For example, solid water, i.e., ice is described as $H_2O(s)$.
- Endothermic reactions are described by adding the term **–Heat** on the product side.
- Exothermic reactions are described by adding the term **+Heat** on the product side.
- Dilute and concentrated solutions are indicated by **dil.** and **conc.** respectively after the formulae of the reactants/products.
- Speed of the reaction can be described by writing **slow** or **fast** over the arrow head.

PARTS OF A CHEMICAL EQUATION

A chemical equation has several parts. Various parts of the following chemical equation are shown below:

$$2HgO(s) \xrightarrow{\Delta} 2Hg(l) + O_2(g)$$

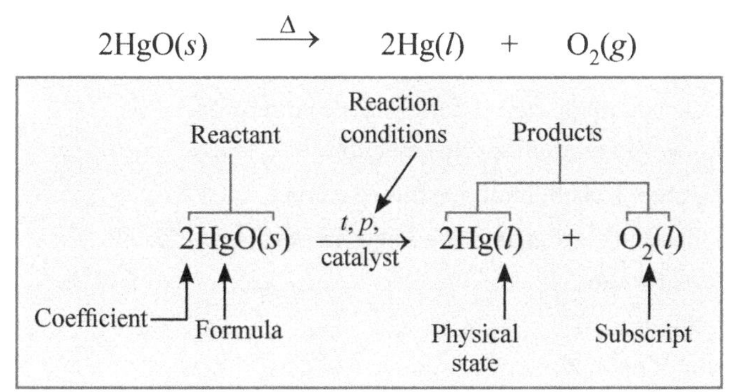

- **Reactants** – The starting substances, which take part in the reaction. (Formulae must be correct).
- **Products** – The substances that are formed in the reaction. (Formulae must be correct).
- **Arrows** – Show the direction in which reaction proceeds.
  - → Used between the reactants and products; means 'react to form'.

- ↛ Used between the reactants and products to show that the equation is not yet balanced.
- ↑ Used with the formula of a gaseous product on its right-hand side.
- ↓ Used with the formula of an insoluble solid product also called a precipitate on its right-hand side.

❖ STATE SYMBOLS – Indicates the physical state of the substances involved in the reaction.

(*g*)	indicates a gaseous substance.
(*l*)	indicates a liquid substance.
(*s*)	indicates a solid substance.
(*aq*)	indicates an aqueous solution.

❖ COEFFICIENTS – The number placed before the symbols or formulae of reactants or products to balance the equation.

❖ CONDITIONS – Words or symbols placed **over** or **under** the horizontal arrow (→) to indicate the reaction conditions.

Δ	indicates that heat is added.
hv	indicates that light is added.
elec	indicates that electrical energy is added.

Self Assessment

1. Write a word equation describing an exothermic reaction.
2. Magnesium reacts with oxygen to form magnesium oxide.
   (a) Write the word equation for this reaction.
   (b) Write the chemical equation for this reaction.
3. How are the following terms described in a chemical equation.
   (a) A solution in water
   (b) Heat is added
   (c) Formation of precipitate of a product.
4. From the statement given below:
   (a) Identify the reactants and products
   (b) Write the chemical equation
   (c) Balance the above chemical equation.
   "Zinc carbonate is heated to form zinc oxide and carbon dioxide."

REVIEW QUESTIONS

A. Very Short Answer Type Questions

1. By what name is the substance which takes part in a reaction called?
2. What are the reactants and products in the following reaction?

$$2H_2(g) + O_2(g) \rightarrow 2H_2O(l)$$

3. Name the reaction in which heat is absorbed.
4. What is given out in an exothermic reaction?
5. Write the word equation for a reaction in which limestone is decomposed to produce lime and carbon dioxide.
6. What is a chemical equation having equal number of atoms of all the elements on both the sides called?
7. What does the arrow drawn downwards (↓) on the right-hand side of a substance in a chemical equation describe?
8. What does the symbol 'Δ' over the arrow indicate in a chemical equation?
9. What does the term *'aq'* written against the formula of a substance in a chemical equation represent?
10. A reaction takes place in the presence of light. How will you provide this information in a chemical equation?

B. Short Answer Type Questions

1. How is a chemical reaction described in Chemistry?
2. What is an endothermic reaction? Give one example.
3. What is a word equation? Why is it not used by Chemists to describe reactions?
4. Write the chemical equation for a reaction in which zinc metal reacts with dilute sulphuric acid to give zinc sulphate and hydrogen gas.
5. What is the process by which the number of atoms of each element on both the sides of a chemical equation is made equal called?

C. Long Answer Type Questions

1. Describe various parts of the chemical equation,

$$2Mg(s) + O_2(g) \xrightarrow{\Delta} 2MgO(s)$$

2. Why is the burning of coal considered an exothermic chemical reaction? Write chemical equation for this reaction.
3. Mention the various steps for writing a chemical equation.
4. How can a chemical equation be made more informative?
5. Write chemical equations for the following chemical reactions:
   (a) Magnesium + Copper(II) sulphate → Copper + Magnesium sulphate
   (b) Sodium carbonate + Hydrochloric acid → Sodium chloride + Carbon dioxide + Water

D. True or False Type Questions

Write 'T' for True or 'F' for False.

1. Substances that take part in a reaction are called the products.
2. Burning of LPG is an endothermic process.
3. Reactants + Heat → Products, describes an exothermic reaction.
4. The chemical reaction described in terms of symbols / formulae of the reactants and products is called a chemical equation.
5. The chemical equation in which atoms of each element on both the sides are equal is called a balanced equation.
6. The reaction conditions are mentioned over the arrow between the reactants and products.
7. The symbol 'Δ' refers to heating during the reaction.
8. The chemical equation, $4Na + O_2 \rightarrow 2Na_2O$ is a balanced equation.
9. In any chemical equation, the symbol ↓ describes the formation of an insoluble substance.
10. The aqueous solution of a substance is described by adding the term (*aq*) after the formula of the substance.

E. Fill in the Blanks Type Questions

1. Burning of magnesium in air is an ______________ reaction.
2. When quicklime is added to water, ______________ is generated.
3. When ammonium chloride is dissolved in water, the solution becomes cold. The process of dissolution of ammonium chloride is an ______________ process.
4. The equation described in terms of names of the reactants and products is called ______________ equation.
5. When atoms of each element on both the sides of a chemical equation are equal, it is called a ______________ chemical equation.
6. The solution of a substance in water is denoted by adding the term ______________ to its formula.
7. The ______________ are written on the left side of the arrow in a chemical equation.
8. The absorption of light by a reactant is denoted by a term ______________ written over the arrow in a chemical equation.
9. A gaseous product formed during a reaction is shown by ______________ in a chemical equation.
10. The chemical equation,

    Reactants + Heat → Products

    is ______________ in nature.

F. Match the Columns Type Questions

a.

	Column A		Column B
1	Exothermic	A.	Substance whose concentration decreases with time
2	Reactant	B.	(*aq*)
3	Chemical equation	C.	↓
4	Solution in water	D.	Symbols and formulae
5	Precipitate	E.	Burning of a fuel

b.

	Column A		Column B
1	Elements	A.	Quicklime
2	Water → Steam	B.	Light
3	Heating	C.	Symbols
4	$h\nu$	D.	Endothermic
5	CaO	E.	Δ

G. Multiple Choice Type Questions

Choose the correct alternative.

1. The substance whose concentration decreases with time during the course of reaction is called
   (a) product ❑ (b) reactant ❑ (c) catalyst ❑ (d) intermediate ❑
2. The process in which heat is evolved is called
   (a) exothermic ❑ (b) reversible ❑ (c) irreversible ❑ (d) endothermic ❑
3. The chemical reaction described by an equation written in terms of the names of the reactants and the products is called
   (a) chemical equation ❑ (b) partial equation ❑
   (c) word equation ❑ (d) none of these ❑
4. Which of the following information is provided over the arrow in a chemical equation?
   (a) temperature, pressure ❑ (b) physical state ❑
   (c) heat evolved or absorbed ❑ (d) all of these ❑
5. The chemical equation, $HgO \rightarrow Hg + O_2$ can be balanced by adding the coefficient ______________ before ______________ and ______________
   (a) 2, HgO, Hg ❑ (b) 2, O_2, Hg ❑ (c) 2, HgO, O_2 ❑ (d) $\frac{1}{2}$, O_2, HgO ❑
6. The kind of chemical equation in the following reaction is

$$Mg + O_2 \rightarrow MgO$$

   (a) balanced ❑ (b) unbalanced ❑
   (c) partially balanced ❑ (d) any of these ❑
7. The heat supplied to a reaction is described by
   (a) Δ over the arrow ❑ (b) Δ under the arrow ❑
   (c) $h\nu$ ❑ (d) elec ❑
8. The formation of precipitate during a chemical reaction is shown by the symbol
   (a) ↑ ❑ (b) ↓ ❑ (c) ↔ ❑ (d) → ❑

ANSWERS

A. 1. Reactant 2. Reactants: $H_2(g)$ and $O_2(g)$; Product: $H_2O(l)$

3. Burning of a fuel 4. Heat 5. Limestone $\xrightarrow{\Delta}$ Lime + Carbon dioxide

6. Balanced equation 7. Formation of precipitate 8. Supplying heat to the reactants

9. Solution of the substance in water 10. Writing *hv* over the arrow

D. 1. F 2. F 3. F 4. T 5. T 6. T 7. T 8. T 9. T 10. T

E. 1. Exothermic 2. Heat 3. Endothermic 4. Word 5. Balanced

6. (*aq*) 7. Reactants 8. *hv* 9. ↑ 10. Endothermic

F. a. 1 – E; 2 – A; 3 – D; 4 – B; 5 – C

b. 1 – C; 2 – D; 3 – E; 4 – B; 5 – A

G. 1. b 2. a 3. c 4. a 5. a 6. b 7. a 8. b

How are new elements named?

IUPAC lets the discoverers of an element propose a name and a two-letter symbol.

As per IUPAC guidelines, a new element only can be named after:

1. a property of the element such as colour (indium, rubidium etc.), smell (osmium and bromine etc.)
2. a mineral or substance e.g. calcium etc.
3. a place - often where they were discovered or synthesized.

   These places range in size from

   - continents e.g. europium
   - countries e.g. americium, francium, polonium
   - villages e.g. ytterbium, yttrium, erbium, terbium, strontium
4. a mythological concept or character e.g. titanium, arsenic, tantalum, nickel and cobalt.
5. a scientist - several scientists have been honoured by having elements named after them, e.g. curium, einsteinium, fermium and mendelevium.

Of course, these names must end in "-ium", "-ine"or "-on" depending on the group they belong to.

Helium is one element that wasn't first discovered on Earth but has been named after its place of the discovery (helios - from the Greek word for Sun)

Seaborgium, was the first element to be named after a living scientist, Glenn Seaborg.

7

Balancing of Chemical Equations

Balancing of chemical equations is based on the **law of conservation of mass.**

The commonly used methods for balancing chemical equations are:

- Hit-and-Trial method
- Partial equation method

> The **Hit-and-Trial** method is also called **Inspection method.** This method is based on **Material balance** or **Mass balance.**

BALANCING BY HIT-AND-TRIAL METHOD

Follow the steps described below for balancing chemical equations by Hit-and-Trial method:

Step 1. Start the process of balancing from the element (other than oxygen and hydrogen) which appears least number of times in the equation.

Step 2. Balance oxygen.

Step 3. Balance hydrogen.

Step 4. In equations, where elementary gases such as H_2, N_2 or O_2 appear as the reactant or the product, these should be kept in the atomic form initially, i.e., as H, N or O.

Step 5. Check, to be sure, that the chemical equation is balanced.

Step 6. Make the equation molecular, if required.

APPLICATION OF HIT-AND-TRIAL METHOD

The application of Hit-and-Trial method for balancing chemical equations is illustrated through the examples given below:

Example 1: Write the chemical equation for the reaction given below and balance it.

$$\text{Magnesium} + \text{Oxygen} \xrightarrow{\text{heat}} \text{Magnesium oxide}$$

Step 1. Convert the given word equation into the skeleton chemical equation.

$$Mg + O_2 \xrightarrow{\text{heat}} MgO$$

Step 2. Count the number of atoms of each element on both the sides of the chemical equation and write in the table.

Element	Number of atoms on the		Remarks
	Left-hand (reactant) side	Right-hand (product) side	
Mg	1	1	Equal
O	2	1	Unequal

Step 3. Oxygen (O) atoms can be balanced by multiplying MgO by two (2). This gives,

$$Mg + O_2 \xrightarrow{\text{heat}} 2MgO$$

Oxygen atoms are balanced.

Step 4. Mg atoms can be balanced by multiplying Mg (on the left) by two (2). This gives,

$$2Mg + O_2 \xrightarrow{\text{heat}} 2MgO$$

The atoms of both Mg and O are now equal on both the sides. Thus, the balanced equation is

$$2Mg + O_2 \xrightarrow{\text{heat}} 2MgO$$

Example 2: Hydrogen and oxygen react to form water. Write the balanced chemical equation for this reaction.

The chemical equation for the reaction between hydrogen and oxygen is written as follows:

Step 1. The reactants and product are identified as:

Reactants: Hydrogen and Oxygen

Product: Water (H_2O)

Step 2. Write formulae of the reactants and product on either side of an arrow. The molecular formula of hydrogen is H_2 and that of oxygen is O_2. So, the skeleton equation is,

$$H_2(g) + O_2(g) \rightarrow H_2O(l)$$

Step 3. Count the number of atoms of each element on both the sides of the chemical equation and write in the following table.

Element	Number of atoms on the		Remarks
	Left-hand (reactant) side	Right-hand (product) side	
H	1	2	Unequal
O	2	1	Unequal

Step 4. The number of atoms of oxygen (O) on both the sides are not equal. The number of oxygen atoms on both the sides can be made equal by placing a coefficient of 2 before H_2O (on the product side). Thus,

$$H_2(g) + O_2(g) \rightarrow 2H_2O(l)$$

Step 5. But, by doing so, the number of hydrogen atoms on the right-hand side has become 4. On the left-hand side, there are only two hydrogen atoms. Hydrogen atoms can be equalised by placing a coefficient of 2 before H_2 (on the reactant side). Then, the chemical equation becomes,

$$2H_2(g) + O_2(g) \rightarrow 2H_2O(l)$$

Step 6. Now, counting of the number of atoms of each element shows that there are four hydrogen atoms and two oxygen atoms on both the sides. Thus, the balanced chemical equation for the reaction between hydrogen and oxygen is,

$$2H_2(g) + O_2(g) \rightarrow 2H_2O(l)$$

Example 3: Write the chemical equation for the following word equation and balance it.

Methane + Oxygen (from air) → Carbon dioxide + Water

Step 1. Convert the given word equation into a chemical equation using symbols and formulae.

$$CH_4(g) + \underset{\text{(from air)}}{O_2(g)} \rightarrow CO_2(g) + H_2O(l)$$

Step 2. Count the number of atoms of each element on both the sides of arrow (→) in the chemical equation and write in the following table.

Element	Number of atoms on the		Remarks
	Left-hand (reactant) side	Right-hand (product) side	
C	1	1	Equal
H	4	2	Unequal
O	2	3	Unequal

Step 3. Inspection of the skeleton equation shows that both carbon (C) and hydrogen (H) occur twice, and oxygen (O) appears thrice. So, a start is made by balancing carbon or hydrogen atoms.

Carbon is already balanced. There are 4 hydrogen atoms on the left-hand side, and 2 hydrogen atoms on the right-hand side. So, hydrogen (H) can be balanced by placing a coefficient of 2 before H_2O on the right-hand side. Then, the resulting chemical equation is,

$$CH_4(g) + O_2(g) \rightarrow CO_2(g) + 2H_2O(l)$$

Step 4. Now, there is one carbon and four hydrogen atoms on either side of the equation. Thus, carbon and hydrogen are balanced. In this partly balanced chemical equation, there are four oxygen atoms on the right-hand side, while there are two on the left-hand side. Oxygen atoms may be balanced by multiplying O_2 on the left-hand side by 2.

Then, the resulting equation is,

$$CH_4(g) + 2O_2(g) \rightarrow CO_2(g) + 2H_2O(l)$$

Step 5. Count the number of atoms of each element on both the sides of the equation and record in the table.

Element	Number of atoms on the		Remarks
	Left-hand (reactant) side	Right-hand (product) side	
Carbon	1	1	Equal
Hydrogen	4	4	Equal
Oxygen	4	4	Equal

The number of atoms of each element on both the sides of the above equation are equal. Therefore, the above chemical equation is a balanced chemical equation.

Example 4: Write the balanced chemical equations for the following chemical reactions:

1. Calcium hydroxide + Carbon dioxide → Calcium carbonate + Water
2. Zinc + Silver nitrate → Zinc nitrate + Silver
3. Lead + Copper chloride → Lead chloride + Copper
4. Barium chloride + Sodium sulphate → Barium sulphate + Sodium chloride

The balanced chemical equations for the given reactions balanced by Hit-and-Trial method are,

1. $Ca(OH)_2$ (calcium hydroxide) + CO_2 (carbon dioxide) → $CaCO_3$ (calcium carbonate) + H_2O (water)
2. Zn (zinc) + $2AgNO_3$ (silver nitrate) → $Zn(NO_3)_2$ (zinc nitrate) + $2Ag$ (silver)
3. Pb (lead) + $CuCl_2$ (copper chloride) → $PbCl_2$ (lead chloride) + Cu (copper)
4. $BaCl_2$ (barium chloride) + Na_2SO_4 (sodium sulphate) → $BaSO_4$ (barium sulphate) + $2NaCl$ (sodium chloride)

Example 5: Write the balanced chemical equations with the state symbols for the following reactions:

1. Iron(III) oxide reacts with carbon to produce iron metal and carbon monoxide.
2. Sodium hydroxide solution (in water) reacts with hydrochloric acid solution (in water) to produce sodium chloride and water.

The balanced chemical equations for the given reactions are,

1. $Fe_2O_3(s)$ (iron(III) oxide) + $3C(s)$ $\xrightarrow{\Delta}$ $2Fe(s)$ + $3CO(g)$
2. $NaOH(aq)$ (sodium hydroxide) + $HCl(aq)$ (hydrochloric acid) → $NaCl(aq)$ (sodium chloride) + $H_2O(l)$ (water)

Example 6: Balance the following unbalanced chemical equation:

$$Fe(s) + H_2O(g) \rightarrow Fe_3O_4(s) + H_2(g)$$

The given unbalanced chemical equation is

$$Fe(s) + H_2O(g) \rightarrow Fe_3O_4(s) + H_2(g)$$

Step 1. Count the number of atoms of all the elements on both the sides of the chemical equation and write in the following table.

Element	Number of atoms on the		Remarks
	Left-hand (reactant) side	Right-hand (product) side	
Fe	1	3	Unequal
H	2	2	Equal
O	1	4	Unequal

Step 2. To balance iron (Fe) atoms, place the coefficient 3 before Fe (on the left).

Step 3. Hydrogen atoms (H) are balanced.

Step 4. To balance the oxygen atoms (O), place the coefficient 4 before H_2O (on the left). The partially balanced equation is,

$$3Fe(s) + 4H_2O(g) \rightarrow Fe_3O_4(s) + H_2(g)$$

Step 5. Now, there are eight hydrogen atoms on the reactant side and only two on the product side. So, to balance H atoms, place the coefficient 4 before H_2 (on the right). Then, we can write,

$$3Fe(s) + 4H_2O(g) \rightarrow Fe_3O_4(s) + 4H_2(g)$$

Step 6. Inspection of this equation shows that atoms of all the elements are equal on both the sides.

Therefore, the above equation is a balanced equation.

Self Assessment 1

1. Name the principle behind the balancing of chemical equations.
2. Name two methods for the balancing of chemical equations.
3. Is the chemical equation given below a balanced or an unbalanced equation?
   $$N_2 + H_2 \rightarrow 2NH_3$$
4. In which chemical equation, the atoms of each element are equal in number on both the sides of the arrow?
5. Can the coefficient before any reactant or product be a fraction?

BALANCING BY PARTIAL EQUATION METHOD

In the partial equation method, the **overall reaction** is assumed to take place through two or more simpler reactions. Each such simpler reaction can be described by a simple chemical equation, called **partial equation.** Each partial equation is balanced by simple Hit-and-Trial method.

To balance a chemical equation by partial equation method, follow the steps given below:

Step 1. Split the given chemical equation into two or more simpler equations called partial equations.

Step 2. Balance each partial equation separately by Hit-and-Trial method.

Step 3. Multiply such balanced partial equations with suitable coefficients so as to exactly cancel out the common substances which do not appear in the overall chemical equation.

Step 4. Add the balanced partial equation obtained in Step 3 to obtain the balanced overall chemical equation.

APPLICATION OF PARTIAL EQUATION METHOD

The use of partial equation method for balancing chemical equations is illustrated below:

Example: Balance the chemical equation given below by partial equation method:

$$KMnO_4 + FeSO_4 + H_2SO_4(aq) \rightarrow K_2SO_4 + MnSO_4 + Fe_2(SO_4)_3 + H_2O$$

Step 1. Write the **skeleton equation** (as given).

$$KMnO_4 + FeSO_4 + H_2SO_4(aq) \rightarrow K_2SO_4 + MnSO_4 + Fe_2(SO_4)_3 + H_2O$$

Step 2. Split the skeleton equation into two simpler equations (called partial equations).

- **Partial Equation 1:**

$$KMnO_4 + H_2SO_4(aq) \rightarrow K_2SO_4 + MnSO_4 + H_2O + O \quad (1)$$

- **Partial Equation 2:**

$$FeSO_4 + H_2SO_4(aq) + O \rightarrow Fe_2(SO_4)_3 + H_2O \quad (2)$$

❖ These partial equations are separately balanced by Hit-and-Trial method as follows:

Balancing Partial Equation 1

$$KMnO_4 + H_2SO_4(aq) \rightarrow K_2SO_4 + MnSO_4 + H_2O + O$$

The number of atoms of different elements which occur in this equation, are

Element	Number of atoms on the		
	Left-hand (reactant) side	Right-hand (product) side	Number of times the element appears
K	1	2	2
Mn	1	1 (equal)	2
H	2	2 (equal)	2
O	8	10	6
S	1	2	3

❖ The elements which appear least are Mn and H. These are already balanced.

❖ **Balancing K atoms** by placing a coefficient of 2 before $KMnO_4$ (on the left-hand side).

- **Balance Mn atoms** by placing a coefficient of 2 before $MnSO_4$ (on the right-hand side). As a result, partial equation 1 becomes,

$$2KMnO_4 + H_2SO_4 \rightarrow K_2SO_4 + 2MnSO_4 + H_2O + O$$

- **Balance S atoms** by placing a coefficient of 3 before H_2SO_4 (on the left-hand side). Then, the above equation becomes,

$$2KMnO_4 + 3H_2SO_4 \rightarrow K_2SO_4 + 2MnSO_4 + H_2O + O$$

- **Balance H atoms** by placing a coefficient of 3 before H_2O (on the right-hand side). Then, the above equation becomes,

$$2KMnO_4 + 3H_2SO_4 \rightarrow K_2SO_4 + 2MnSO_4 + 3H_2O + O$$

- **Balance O atoms** by placing a coefficient of 5 before the symbol of O (on the right-hand side). Then, the balanced partial equation 1 becomes,

$$2KMnO_4 + 3H_2SO_4 \rightarrow K_2SO_4 + 2MnSO_4 + 3H_2O + 5O \quad \textbf{(a)}$$

Balancing Partial Equation 2

$$FeSO_4 + H_2SO_4 + O \rightarrow Fe_2(SO_4)_3 + H_2O$$

- The elements which appear in the equation least are H and Fe. Here, H atoms on the two sides are already equal. So, balancing is started with Fe.

- **Balance Fe atoms** by placing a coefficient of 2 before $FeSO_4$ (on the left-hand side). Then, the partial equation 2 becomes,

$$2FeSO_4 + H_2SO_4 + O \rightarrow Fe_2(SO_4)_3 + H_2O \quad \textbf{(b)}$$

This partial equation is balanced.

Step 3. The two balanced partial equations are then balanced with respect to each other as follows:

Balancing the Balanced Partial Equations w.r.t. Each other

- Balance the two equations **(a)** and **(b)** with respect to each other by multiplying equation **(b)** with a coefficient of 5 and adding to equation **(a)**.
  - Balanced partial equation **(a)**

$$2KMnO_4 + 3H_2SO_4 \rightarrow K_2SO_4 + 2MnSO_4 + 3H_2O + 5O$$

  - Balanced partial equation **(b)** × 5

$$2FeSO_4 + H_2SO_4 + O \rightarrow Fe_2(SO_4)_3 + H_2O \times 5$$

  - Add the balanced equations **(a)** and **(b)**. The overall balanced equation is

$$2KMnO_4 + 10FeSO_4 + 8H_2SO_4 \rightarrow K_2SO_4 + 2MnSO_4 + 5Fe_2(SO_4)_3 + 8H_2O$$

Self Assessment 2

1. A chemical reaction involves many reactants. Which method of balancing will be used for balancing the chemical equation for such a reaction?
2. Why is the chemical equation for a reaction involving many reactants split into two or more simpler equations?
3. By what name are the simpler equations called?
4. Name the smallest unit of matter that takes part in a chemical reaction.
5. Do atoms retain their identity during chemical reactions?

REVIEW QUESTIONS

A. Very Short Answer Type Questions

1. What is the chemical equation in which atoms of each element on both the sides are equal called?
2. Which law governs the balancing of a chemical equation?
3. Name the simplest method of balancing chemical equations.
4. Name the method of balancing chemical equations by splitting the overall chemical equation into two or more simpler equations.
5. State the law of conservation of mass.

B. Short Answer Type Questions

1. Describe in brief, the Hit-and-Trial method of balancing chemical equations.
2. What is meant by mass balance?
3. Balance the chemical equation by Hit-and-Trial method:

$$Na + O_2 \xrightarrow{\Delta} Na_2O$$

4. Write a balanced chemical equation for the following reaction. Aluminium is oxidised by oxygen to form aluminium oxide.
5. Write the balanced chemical equation for a reaction involving:

   Reactants: Sodium metal, Water

   Products: Sodium hydroxide, Hydrogen

C. Long Answer Type Questions

1. Describe the various steps involved in balancing of a chemical equation by Hit-and-Trial method.
2. Describe the partial equation method of balancing chemical equations.
3. What is meant by partial equations? Illustrate by giving a simple example.
4. From the statement given below:

   "Carbon dioxide passed through limewater gives white turbidity due to the formation of calcium carbonate".

   (a) Identify the reactants and products.

   (b) Write the chemical equation (including state symbols).

   (c) Balance the chemical equation.

5. Provide the following information from the following word equation:
   Sulphur dioxide + Chlorine + Water → Sulphuric acid + Hydrochloric acid
   (a) Write the formulae of the reactants and products.
   (b) Rewrite the given word equation as the balanced chemical equation (including state symbols).

D. True or False Type Questions

Write 'T' for True or 'F' for False.

1. A balanced chemical equation involves equal number of atoms of each element on both the sides of the chemical equation.
2. The Hit-and-Trial method of balancing chemical equations is based on charge balance.
3. The number placed before the formula of any reactant or product is called stoichiometric coefficient.
4. In partial equation method, the overall reaction is assumed to take place through two or more simpler reactions.
5. The stoichiometric coefficients in a chemical equation can be written as a fraction.

E. Fill in the Blanks Type Questions

1. NH_3 + ________ → NH_4NO_3
2. Fe + $HCl(aq)$ → ________ + H_2
3. ________ + $HCl(aq)$ → NaCl + H_2O
4. Zn + $H_2SO_4(aq)$ → ________ + $H_2(g)$
5. $KOH(aq)$ + $HCl(aq)$ → ________ + $H_2O(l)$
6. $Pb(NO_3)_2(aq)$ + ________ → $PbI_2(s)$ + ________
7. ________ $\xrightarrow{\Delta}$ $ZnO(s)$ + $CO_2(g)$
8. $ZnCO_3$ + ________ → $ZnSO_4$ + $CO_2(g)$ + H_2O
9. KI + ________ + H_2O → KOH + I_2 + O_2
10. SO_2 + ________ → S + H_2O

F. Match the Columns Type Questions

a.

	Column A		Column B
1	Mass balance	A.	Consumed
2	Reactant	B.	Produced
3	Unbalanced	C.	Conservation of mass
4	Product	D.	Simpler reaction
5	Partial equation	E.	Chemical equation

b.

	Column A		Column B
1	$CoBr_2$	A.	Lead sulphide
2	IO_2	B.	Silicon dioxide
3	PbS	C.	Phosphorus
4	SiO_2	D.	Cobalt(II) bromide
5	P_4	E.	Iodine dioxide

G. Multiple Choice Type Questions

Choose the correct alternative.

1. The Hit-and-Trial method of balancing chemical equations is based on
   (a) number balance ❑ (b) mass balance ❑
   (c) charge balance ❑ (d) all of these ❑
2. The chemical equation, $H_2 + O_2 \rightarrow 2H_2O$ can be balanced by placing
   (a) 2 before H_2 ❑ (b) $\frac{1}{2}$ before O_2 ❑
   (c) $\frac{1}{2}$ before H_2O ❑ (d) 4 before H_2 ❑
3. Identify the major reactant in the reaction,
   $4Na + O_2 \rightarrow 2Na_2O$
   (a) O_2 ❑ (b) Na_2O ❑
   (c) Na ❑ (d) all of these ❑
4. The stoichiometric coefficients in a balanced chemical equation are always
   (a) integral number ❑ (b) half-integral ❑
   (c) negative number ❑ (d) any of these ❑
5. Which of the following can be balanced?
   (a) word equation ❑ (b) mathematical equation ❑
   (c) chemical equation ❑ (d) none of these ❑

ANSWERS

A. 1. Balanced equation 2. Law of conservation of mass 3. Hit-and-Trial method
4. Partial equation method 5. Matter (hence mass) can neither be created nor destroyed.

D. 1. T 2. F 3. T 4. T 5. F

E. 1. HNO_3 2. $FeCl_2$ 3. NaOH 4. $ZnSO_4$ 5. KCl
6. KI 7. $ZnCO_3$ 8. H_2SO_4 9. O_3 10. H_2S

F. a. 1 – C; 2 – A; 3 – E; 4 – B; 5 – D
b. 1 – D; 2 – E; 3 – A; 4 – B; 5 – C

G. 1. b 2. a 3. c 4. a 5. c

Balancing of Ionic Equations

You have learnt about the chemical equations. In these equations, the reactants and products are described by atoms and molecules. In most cases, the reactants and products are present in the solution form. In such solutions, the electrolytes dissociate or ionise to give free ions. These free ions then take part in the reaction. Therefore, it is appropriate to describe such reactions by **ionic equations** also.

IONIC EQUATION – The chemical equation described in terms of ions and molecules is called ionic equation.

NET IONIC EQUATION – The ionic equation from which common ions are cancelled out and removed is called net ionic equation.

WRITING IONIC EQUATIONS

To write an ionic equation, follow the steps given below:

Step 1. Write the chemical equation for the given reaction.

Step 2. Write the correct physical states for all the reactants and products.

Step 3. Describe the soluble electrolytes in terms of their ions.

Step 4. Denote the insoluble and sparingly soluble substances in the molecular form.

Step 5. Cancel out the common ions that occur on both the reactant and the product sides.

Step 6. Write the chemical equation in terms of the remaining ions and molecules. This is called the **net ionic equation.**

Step 7. Balance the net ionic equation both in terms of **mass balance** and **charge balance,** if required.

> The ions that do not participate directly in the reaction are called **spectator ions.**

Example 1: Write the net ionic equation for the reaction between silver nitrate solution and sodium chloride solution.

Proceed as follows:

Step 1. Write the word equation and the chemical equation for the reaction between silver nitrate solution and sodium chloride solution.

Silver nitrate soln. + Sodium chloride soln. → Silver chloride (ppt.) + Sodium nitrate soln.

$$AgNO_3(aq) + NaCl(aq) \rightarrow AgCl(s) + NaNO_3(aq)$$

Step 2. Describe the soluble electrolytes in terms of their constituent ions.

$$Ag^+(aq) + NO_3^-(aq) + Na^+(aq) + Cl^-(aq) \rightarrow AgCl(s)\downarrow + Na^+(aq) + NO_3^-(aq)$$

This is the **ionic equation** for the reaction between silver nitrate and sodium chloride.

Step 3. Cancel the common ions from the reactant and product sides of the ionic equation.

$$Ag^+(aq) + \cancel{NO_3^-(aq)} + \cancel{Na^+(aq)} + Cl^-(aq) \rightarrow AgCl(s)\downarrow + \cancel{Na^+(aq)} + \cancel{NO_3^-(aq)}$$

Step 4. Write the **net ionic equation** for the reaction.

$$Ag^+(aq) + Cl^-(aq) \rightarrow AgCl(s)$$

This net ionic equation is balanced both in terms of mass and charge.

Example 2: Write the net ionic equation for the reaction between barium chloride ($BaCl_2$) and lead nitrate [$Pb(NO_3)_2$] solutions.

Step 1. Write the skeleton chemical equation for the given reaction.

$$BaCl_2(aq) + Pb(NO_3)_2(aq) \rightarrow PbCl_2(s)\downarrow + Ba(NO_3)_2(aq)$$

Step 2. Write the **ionic equation.**

$$Ba^{2+}(aq) + 2Cl^-(aq) + Pb^{2+}(aq) + 2NO_3^-(aq) \rightarrow PbCl_2(s)\downarrow + Ba^{2+}(aq) + 2NO_3^-(aq)$$

Step 3. Cancel the common ions from the reactant and product sides of the ionic equation.

$$\cancel{Ba^{2+}(aq)} + 2Cl^-(aq) + Pb^{2+}(aq) + \cancel{2NO_3^-(aq)} \rightarrow PbCl_2(s)\downarrow + \cancel{Ba^{2+}(aq)} + \cancel{2NO_3^-(aq)}$$

Step 4. Write the **net ionic equation** for the reaction.

$$Pb^{2+}(aq) + 2Cl^-(aq) \rightarrow PbCl_2(s)$$

This net ionic equation is balanced both in terms of mass and charge.

Example 3: What will happen if the aqueous solutions of $CaBr_2$ and $(NH_4)_2CO_3$ are mixed? Write the net ionic equation for the reaction.

Follow the steps described below:

Step 1. Write the skeleton chemical equation for the given reaction.

$$CaBr_2(aq) + (NH_4)_2CO_3(aq) \rightarrow CaCO_3 + 2NH_4Br$$

Step 2. From the solubility data, it is found that NH_4Br is **soluble** in water, whereas $CaCO_3$ is sparingly soluble in water. The completed chemical equation is,

$$CaBr_2(aq) + (NH_4)_2CO_3(aq) \rightarrow CaCO_3(s)\downarrow + 2NH_4Br(aq)$$

Step 3. Write the ionic equation.

$$Ca^{2+}(aq) + 2Br^-(aq) + 2NH_4^+(aq) + CO_3^{2-}(aq) \rightarrow CaCO_3(s)\downarrow + 2NH_4^+(aq) + 2Br^-(aq)$$

Step 4. Cancel the common ions from the reactant and product sides of the ionic equation.

$$Ca^{2+}(aq) + \cancel{2Br^-(aq)} + \cancel{2NH_4^+(aq)} + CO_3^{2-}(aq) \rightarrow CaCO_3(s)\downarrow + \cancel{2NH_4^+(aq)} + \cancel{2Br^-(aq)}$$

Step 5. Write the **net ionic equation** for the reaction.

$$Ca^{2+}(aq) + CO_3^{2-}(aq) \rightarrow CaCO_3(s)\downarrow$$

In this reaction, $CaCO_3(s)$ gets precipitated and the solution contains NH_4^+ and Br^- ions only.

Self Assessment

1. What is the equation described in terms of ions and molecules called?
2. What specific information is required to write an ionic equation?
3. What are the ions that do not take part directly in the reaction called?
4. By which method, an ionic equation is balanced?

REVIEW QUESTIONS

A. Very Short Answer Type Questions

1. How is an insoluble substance described in an ionic equation?
2. How is a soluble electrolyte described in an ionic equation?
3. Silver chloride (AgCl) is a sparingly soluble electrolyte. Write the ionic equation for its dissolution.
4. Name the method of balancing an ionic equation.
5. Which ions in the ionic equation do not take part in the reaction?

B. Short Answer Type Questions

1. What is an ionic equation? Give one example.
2. Name two conditions which must be satisfied by an ionic equation.
3. Write the ionic equation for the reaction between the solution of Na_3PO_4 and $CaCl_2$. Identify the spectator ions.
4. What happens when aqueous solutions of aluminium nitrate and barium chloride are mixed? Write the balanced
   (a) chemical equation, and (b) net ionic equation.
5. What happens when aqueous solutions of $(NH_4)_2S$ and $Cu(NO_3)_2$ are mixed? Write the balanced
   (a) chemical equation, and (b) net ionic equation.

C. Long Answer Type Questions

1. How is an ionic equation written?
2. Name the two parameters which must be satisfied during balancing of an ionic equation.
3. Write the ionic equations for the following word equations:
   (a) Ammonium chloride + Sodium hydroxide → Ammonia + Water + Sodium chloride
   (b) Silver nitrate + Potassium iodide → Silver iodide (solid) + Potassium nitrate

4. For each of the following mixtures of the two aqueous solutions,
   (a) Write the balanced chemical equation.
   (b) Write the ionic equation.
   (c) Write the net ionic equation.
   (d) Indicate spectator ions in each case, if any

(i)	$BaCl_2$	and	Na_2SO_4	(v)	KBr	and	$MgCl_2$
(ii)	$AgNO_3$	and	KI	(vi)	Na_3PO_4	and	$CaCl_2$
(iii)	$AlBr_3$	and	LiOH	(vii)	$Ca(NO_3)_2$	and	LiF
(iv)	$Fe(NO_3)_2$	and	$(NH_4)_2CO_3$	(viii)	$Pb(NO_3)_2$	and	NaI

5. For each of the following pairs of the reactants (in solution form), write
   (a) chemical equation, and
   (b) net ionic equation.
      (i) Potassium phosphate and Calcium nitrate
      (ii) Barium chloride and Sodium sulphate
      (iii) Sodium acetate and Potassium nitrate
      (iv) Hydrochloric acid and Lead acetate
      (v) Barium nitrate and Ammonium iodide
      (vi) Potassium hydroxide and Copper(II) nitrate
      (vii) Calcium iodide and Potassium carbonate
      (viii) Magnesium acetate and Barium hydroxide
      (ix) Ammonium bromide and Sodium hydroxide
      (x) Sulphuric acid and Potassium hydroxide
      (xi) Calcium carbonate and Hydrobromic acid
      (xii) Ammonium bromide and Barium acetate
      (xiii) Barium hydroxide and Ammonium bromide

D. True or False Type Questions

Write 'T' for True or 'F' for False.

1. Ionic equations are the chemical equations described in terms of ions only.
2. While balancing an ionic equation, both mass and charge are balanced.
3. The chemical equation, $Ag^+(aq) + Cl^-(aq) \rightarrow AgCl(s)$ is an ionic equation.
4. The ions which do not take part in the reaction are called spectator ions.
5. The reactions involving nonelectrolytes can also be described by net ionic equations.

E. Fill in the Blanks Type Questions

Complete the following chemical equations and
   (a) Write the corresponding chemical equations.
   (b) Balance the chemical equations.
   (c) Convert the balanced chemical equations into ionic equations.

1. $\_\_\_\_\_\_\_\_ + HCl(aq) \rightarrow FeCl_2(aq) + H_2(g)$
2. $Mg(OH)_2(aq) + \_\_\_\_\_\_\_\_ \rightarrow MgSO_4(aq) + H_2O(l)$

3. ______ + $HCl(aq)$ → $NH_4Cl(aq)$
4. $CuO(s)$ + $H_2SO_4(aq)$ → ______ + $H_2O(l)$
5. $NH_4OH(aq)$ + ______ → $(NH_4)_2SO_4(aq)$ + $H_2O(l)$

F. Match the Columns Type Questions

a.

	Column A		Column B
1	$Na_2CO_3(aq)$ + $2HCl(aq)$	A.	$NaCl(aq)$ + $H_2O(l)$ + $CO_2(g)$
2	$Na_2CO_3(aq)$ + $H_2SO_4(aq)$	B.	$Na_2SO_4(aq)$ + $2H_2O(l)$
3	$NaHCO_3(s)$ + $HCl(aq)$	C.	$NaCl(aq)$ + $H_2O(l)$
4	$2NaOH(aq)$ + $H_2SO_4(aq)$	D.	$Na_2SO_4(aq)$ + $CO_2(g)$ + $H_2O(l)$
5	$NaOH(aq)$ + $HCl(aq)$	E.	$2NaCl(aq)$ + $CO_2(g)$ + $H_2O(l)$

b.

	Column A		Column B
1	$Zn(OH)_2(s)$ + $2HCl(aq)$	A.	$ZnCl_2(aq)$ + $H_2O(l)$
2	$ZnO(s)$ + $H_2SO_4(aq)$	B.	$ZnSO_4(aq)$ + $H_2O(l)$ + $CO_2(g)$
3	$ZnCO_3(s)$ + $2HCl(aq)$	C.	$ZnSO_4(aq)$ + $H_2O(l)$
4	$ZnO(s)$ + $2HCl(aq)$	D.	$ZnCl_2(aq)$ + $H_2O(l)$ + $CO_2(g)$
5	$ZnCO_3(s)$ + $H_2SO_4(aq)$	E.	$ZnCl_2(aq)$ + $2H_2O(l)$

G. Multiple Choice Type Questions

Choose the correct alternative.

1. The substances which dissociate/ionise into ions when dissolved in water are called
   (a) nonelectrolytes ☐ (b) electrolytes ☐
   (c) molecular compounds ☐ (d) network solids ☐
2. The chemical equation described in terms of ions is called
   (a) balanced chemical equation ☐ (b) ionic equation ☐
   (c) molecular equation ☐ (d) none of these ☐
3. Which of the following is/are spectator ion/ions in the following reaction:
   $$Na_2CO_3 + H_2SO_4 \rightarrow Na_2SO_4 + H_2O + CO_2$$
   (a) Na^+ ☐ (b) SO_4^{2-} ☐
   (c) Na^+ and SO_4^{2-} ☐ (d) none of these ☐
4. The reacting ions in the reaction,
   $$Pb(NO_3)_2(aq) + 2NaI(aq) \rightarrow PbI_2(s)\downarrow + 2NaNO_3(aq)$$
   are
   (a) Na^+ and NO_3^- ☐ (b) Pb^{2+} and NO_3^- ☐
   (c) Pb^{2+} and I^- ☐ (d) Na^+ and I^- ☐

ANSWERS

A. 1. Molecular form 2. Ionic form 3. $AgCl(s) \rightarrow Ag^+(aq) + Cl^-(aq)$

4. Hit-and-Trial method 5. Spectator

D. 1. F 2. T 3. T 4. T 5. F

E. 1. (a) $Fe(s) + HCl(aq) \rightarrow FeCl_2(aq) + H_2(g)$

(b) $Fe(s) + 2HCl(aq) \rightarrow FeCl_2(aq) + H_2(g)$

(c) $Fe(s) + 2H^+(aq) + 2Cl^-(aq) \rightarrow Fe^{2+}(aq) + 2Cl^-(aq) + H_2(g)$

or $Fe(s) + 2H^+(aq) \rightarrow Fe^{2+}(aq) + H_2(g)$

2. (a) $Mg(OH)_2 + H_2SO_4(aq) \rightarrow MgSO_4(aq) + H_2O(l)$

(b) $Mg(OH)_2(aq) + H_2SO_4(aq) \rightarrow MgSO_4(aq) + 2H_2O(l)$

(c) $Mg^{2+}(aq) + 2OH^-(aq) + 2H^+(aq) + SO_4^{2-}(aq)$

$\rightarrow Mg^{2+}(aq) + SO_4^{2-}(aq) + 2H_2O(l)$

or $2OH^-(aq) + 2H^+(aq) \rightarrow 2H_2O(l)$

3. (a) $NH_3(aq) + HCl(aq) \rightarrow NH_4Cl(aq)$

(b) Equation (a) is balanced.

(c) $NH_3(aq) + H^+(aq) + Cl^-(aq) \rightarrow NH_4^+(aq) + Cl^-(aq)$

or $NH_3(aq) + H^+(aq) \rightarrow NH_4^+(aq)$

4. (a) $CuO(s) + H_2SO_4(aq) \rightarrow CuSO_4(aq) + H_2O(l)$

(b) Equation (a) is balanced.

(c) $CuO(s) + 2H^+(aq) + SO_4^{2-}(aq) \rightarrow Cu^{2+}(aq) + SO_4^{2-}(aq) + H_2O(l)$

or $CuO(s) + 2H^+(aq) \rightarrow Cu^{2+}(aq) + H_2O(l)$

5. (a) $NH_4OH(aq) + H_2SO_4(aq) \rightarrow (NH_4)_2SO_4(aq) + H_2O$

(b) $2NH_4OH(aq) + H_2SO_4(aq) \rightarrow (NH_4)_2SO_4(aq) + 2H_2O$

(c) $2NH_4^+ + 2OH^- + 2H^+ + SO_4^{2-} \rightarrow 2NH_4^+(aq) + SO_4^{2-}(aq) + 2H_2O(l)$

or $2OH^-(aq) + 2H^+(aq) \rightarrow 2H_2O(l)$

F. a. 1 – A; 2 – D; 3 – E; 4 – B; 5 – C

b. 1 – A; 2 – C; 3 – D; 4 – E; 5 – B

G. 1. b 2. b 3. c 4. c

Oxidation, Reduction and Redox Reactions

Oxidation Reaction – Any process which involves the addition of oxygen or any other electronegative radical, or the removal of hydrogen or any other electropositive radical is termed as **oxidation.**

- **Addition of oxygen**

oxidation (addition of oxygen)

$$\underset{\text{carbon}}{C(s)} + \underset{\text{oxygen}}{O_2(g)} \xrightarrow{\text{burning}} \underset{\text{carbon dioxide}}{CO_2(g)}$$

- **Removal of hydrogen**

oxidation (removal of hydrogen)

$$\underset{\text{hydrogen sulphide}}{H_2S(aq)} + \underset{\text{bromine}}{Br_2(aq)} \rightarrow \underset{\text{hydrobromic acid}}{2HBr(aq)} + \underset{\text{sulphur}}{S}$$

- **Addition of an electronegative radical**

oxidation (addition of electronegative radical Cl^-)

$$\underset{\text{iron(II) chloride}}{2FeCl_2(aq)} + \underset{\text{chlorine}}{Cl_2(g)} \rightarrow \underset{\text{iron(III) chloride}}{2FeCl_3(aq)}$$

- **Removal of an electropositive radical**

oxidation (removal of electropositive radical K^+)

$$\underset{\text{potassium iodide}}{2KI(aq)} + \underset{\text{hydrogen peroxide}}{H_2O_2(aq)} \rightarrow \underset{\text{iodine}}{I_2(aq)} + \underset{\text{potassium hydroxide}}{2KOH(aq)}$$

Reduction Reaction – Any process which involves the addition of hydrogen or any other electropositive radical, or the removal of oxygen or any other electronegative radical is termed as **reduction.**

- **Addition of hydrogen**

reduction (addition of hydrogen)

$$\underset{\text{hydrogen sulphide}}{H_2S(aq)} + \underset{\text{chlorine}}{Cl_2(g)} \rightarrow \underset{\text{hydrochloric acid}}{2HCl(aq)} + \underset{\text{sulphur}}{S(s)}$$

- **Removal of oxygen**

reduction (removal of oxygen)

$$\underset{\text{copper(II) oxide}}{CuO(s)} + \underset{\text{hydrogen}}{H_2(g)} \xrightarrow{\Delta} \underset{\text{copper}}{Cu(aq)} + \underset{\text{water}}{H_2O(l)}$$

- **Addition of an electropositive radical**

reduction (addition of electropositive radical Fe^{2+})

$$\underset{\text{ferric chloride}}{2FeCl_3(aq)} + \underset{\text{iron}}{Fe(s)} \rightarrow \underset{\text{ferrous chloride}}{3FeCl_2(aq)}$$

- **Removal of an electronegative radical**

reduction (removal of electronegative radical Cl^-)

$$\underset{\text{ferric chloride}}{2FeCl_3(aq)} + \underset{\text{hydrogen sulphide}}{H_2S(g)} \rightarrow \underset{\text{ferrous chloride}}{2FeCl_2(aq)} + \underset{\text{hydrochloric acid}}{2HCl(aq)} + \underset{\text{sulphur}}{S(s)}$$

OXIDISING AGENT – A substance which can bring about oxidation of other substances is called an **oxidising agent.**

In other words, a substance which causes the addition of oxygen or removal of hydrogen from other substances is called an **oxidising agent.**

In the reaction,

$$H_2S(aq) + Cl_2(g) \rightarrow 2HCl(aq) + S(s)$$

Chlorine is the **oxidising agent.**

Some common oxidising agents are:

- Oxygen
- Ozone
- Hydrogen peroxide
- Chlorine
- Nitric acid (conc.)
- Sulphuric acid (conc.)
- Potassium permanganate
- Potassium dichromate.

Fluorine is the strongest oxidising agent because of its highest electronegativity.

REDUCING AGENT – A substance which can bring about reduction of other substances is called a **reducing agent.**

In other words, a substance which causes the addition of hydrogen or removal of oxygen from other substances is called a **reducing agent.**

In the reaction,

$$\underset{\text{carbon}}{C(s)} + \underset{\text{zinc oxide}}{ZnO(s)} \xrightarrow{\text{heat}} \underset{\text{carbon monoxide}}{CO(g)} + \underset{\text{zinc}}{Zn(s)}$$

Carbon is the **reducing agent.**

Some common reducing agents are:

- Hydrogen
- Carbon
- Sulphur dioxide
- Hydrogen sulphide.

ELECTRONIC CONCEPT OF OXIDATION AND REDUCTION

According to the electronic concept, the oxidation and reduction are described in terms of loss or gain of electrons by any chemical species. These are defined below:

OXIDATION – The process which involves a **loss of electrons,** is called oxidation.

In the reaction,

$$\underset{\text{copper atom}}{Cu} \rightarrow \underset{\text{copper(II) ion}}{Cu^{2+}} + \underset{\text{electrons}}{2e^-}$$

Copper atom (Cu) loses two electrons and gets oxidised to Cu^{2+} ion.

REDUCTION – The process which involves the **gain of electrons,** is called reduction. In the reaction,

$$\underset{\text{iron(III) ion}}{Fe^{3+}} + \underset{\text{electron}}{e^-} \rightarrow \underset{\text{iron(II) ion}}{Fe^{2+}}$$

Iron(III) ion (also called **ferric ion)** gains one electron and gets reduced to iron(II) ion (also called **ferrous ion).**

OXIDISING AGENT – A chemical species which can pull out/accept electrons from other species is called an **oxidising agent.**

REDUCING AGENT – A chemical species which can give electrons to some other species is called a **reducing agent.**

OXIDATION NUMBER – The number of electrons lost or gained or shifted partially away or towards the atom of an element during a chemical reaction is called its **oxidation number** (O.N.).

- When electrons are **lost** or **shifted partially away** from an atom, its oxidation number is said to be **positive.**
- When electrons are **gained** or **shifted towards** an atom during a reaction, its oxidation number is said to be **negative.**

Oxidation number is also known as oxidation state.

COUNTING OF ELECTRONS DURING THE REMOVAL OF AN ATOM FROM ANY CHEMICAL SPECIES

During the removal of atoms as ions from any chemical species, the electrons on the atoms are counted according to the following rules:

Rule 1. The electrons shared between two **like atoms** are divided **equally** between the sharing atoms. In such a case, the oxidation number of the element in its molecule is **zero.**

For example, in a hydrogen molecule (H_2), the shared pair of electrons is equally shared by the two atoms (H : H). As a result, none of the H-atoms acquires any electrical charge. Therefore, **the oxidation number of hydrogen in hydrogen molecule (H_2) is zero (0).**

In general, all elements in their elemental form have zero oxidation number.

Rule 2. The electrons shared between two **unlike atoms** are counted with **more electronegative** atom. Thus, when such an atom is removed from any species, it is considered to go out as an **anion.**

For example, the shared pair of electrons in a molecule of hydrogen chloride (HCl) is attracted towards more electronegative chlorine atom. As a result, when Cl atom is removed from HCl, it goes out as chloride ion (Cl^-). Therefore,

The oxidation number of Cl in HCl is –1. This leaves hydrogen atom of HCl with one positive charge. Therefore, the **oxidation number of H in HCl is +1.**

RULES FOR CALCULATING THE OXIDATION NUMBER OF AN ELEMENT IN ANY CHEMICAL SPECIES

The following rules are used to determine the oxidation number of an element in any chemical species:

Rule 1. An element in its elemental form has an oxidation number of zero (0).

For example, the oxidation number of hydrogen in H_2, oxygen in O_2, chlorine in Cl_2, phosphorus in P_4, sulphur in S_8, iron in Fe, sodium in Na, etc., is zero. The oxidation number of helium in He is also zero.

Rule 2. The oxidation number of hydrogen is +1 per atom except in the case of metal hydrides.

In metal hydrides such as LiH, NaH, CaH_2, MgH_2, etc., hydrogen has an oxidation number of –1.

Rule 3. The oxidation number of oxygen in the combined form is –2 (except in peroxides and oxyfluorides).

In **peroxides,** the oxidation number of oxygen is –1 and in **oxyfluorides,** the oxidation number of oxygen is +2.

- The oxidation number of oxygen in sodium peroxide (Na_2O_2) and hydrogen peroxide (H_2O_2) is –1.
- In the case of OF_2, oxygen has an oxidation number of +2.

Rule 4. The oxidation number of fluorine (the most electronegative element) in the combined form is –1.

For other halogens, the oxidation number is always –1, except when bonded to a more electronegative halogen or oxygen.

Chlorine, Bromine and Iodine form oxides in which the oxidation number of halogen ranges from +1 to +7.

Oxidation no. of halogen	:	+1	+2	+3	+4	+5	+6	+7
Halogen oxide	:	Cl_2O	–	Cl_2O_3	ClO_2, BrO_2	I_2O_5	Cl_2O_6	Cl_2O_7

Rule 5. The oxidation number of an element in its monatomic ion is equal to the charge on the ion.

The oxidation number of Na in Na^+ and of K in K^+ is +1.

The oxidation number of chlorine in Cl^- is –1.

Ba^{2+} and Al^{3+} have oxidation numbers of +2 and +3 respectively.

Rule 6. The algebraic sum of the oxidation numbers of all the atoms in a neutral molecule is equal to zero (0).

Rule 7. For **polyatomic ions** such as ClO_3^-, PO_4^{3-}, SO_4^{2-}, etc., the **sum of the oxidation numbers** of all the atoms in them is equal to the **net charge** on the ion.

The assignment of oxidation numbers of the atoms in covalent compounds and polyatomic ions (or radicals) is a formality that is helpful in balancing the redox equations. However, the oxidation number of any atom in such compounds **does not necessarily** describe its electronic state in the molecule.

CALCULATION OF OXIDATION NUMBER OF AN ELEMENT IN A CHEMICAL SPECIES

Calculate the oxidation number of the following:

- Mn in $KMnO_4$
- Cr in $K_2Cr_2O_7$
- S in $Na_2S_2O_3$.

Oxidation number of Mn in $KMnO_4$

$KMnO_4$ contains elements K, Mn and O. From the rules given above, the oxidation number of K is +1 and that of O is –2. The molecule $KMnO_4$ has no charge on it.

Let the oxidation number of Mn be x.

Then, $$1 + x + 4 \times (-2) = 0$$

or $$1 + x - 8 = 0$$

or $$x = +7$$

Therefore, the oxidation number of Mn in $KMnO_4$ is +7.

Oxidation number of Cr in $K_2Cr_2O_7$

$K_2Cr_2O_7$ contains elements K, Cr and O. From the rules give above the oxidation number of K is +1 and that of O is –2. The molecule $K_2Cr_2O_7$ has no charge on it.

Let the oxidation number of Cr be x.

Then, $2 \times (+1) + 2 \times x + 7 \times (-2) = 0$

or $2x = 14 - 2 = +12$

or $x = +6$

Therefore, the oxidation number of Cr in $K_2Cr_2O_7$ is +6.

Oxidation number of S in $Na_2S_2O_3$

$Na_2S_2O_3$ contains elements Na, S and O. As per rules given above, the oxidation number of Na is +1 and that of O is –2. The molecule $Na_2S_2O_3$ has no charge on it.

Let the oxidation number of S be x.

Then, $2 \times 1 + 2 \times x + 3 \times (-2) = 0$

or $2 + 2x - 6 = 0$

or $x = +2$

Therefore, the oxidation number of S in $Na_2S_2O_3$ is +2.

OXIDATION NUMBER AND NOMENCLATURE

Some metals form two monatomic cations exhibiting different oxidation states. These oxidation states are distinguished by using the suffixes **-ous** and **-ic.**

- The suffix **-ous** is used for the cation having lower oxidation number.
- The suffix **-ic** is used for the cation having higher oxidation number.

Examples:

- Copper in the combined state occurs as Cu^+ and Cu^{2+}.
- The oxidation number of copper in Cu^+ is +1, while that in Cu^{2+} is +2.

Thus,

- Cu^+ is called **cuprous ion,** while Cu^{2+} is called **cupric ion.**

The names of certain most common cations showing two different oxidation numbers are given below:

Tin		Iron		Mercury	
Sn^{2+}	Sn^{4+}	Fe^{2+}	Fe^{3+}	Hg^+	Hg^{2+}
Stannous	Stannic	Ferrous	Ferric	Mercurous	Mercuric

This method, however, is not applicable to the metals which exhibit more than two oxidation numbers. For such cases, the cations having different oxidation numbers are distinguished by using **Stock notation.**

Stock notation is not used for nonmetals.
Compounds such as PCl_3 and PCl_5 are distinguished by naming them as phosphorus trichloride and phosphorus pentachloride respectively.

THE STOCK NOTATION

According to the **Stock notation,** the oxidation number of a metal is indicated by a **Roman numeral** enclosed in parentheses and written just after the symbol/name of the metal. For example,

Cu^+	is indicated as Cu(I)	So, Cu_2O	is written as	Copper(I) oxide
Cu^{2+}	is indicated as Cu(II)	So, CuO	is written as	Copper(II) oxide
Cr^{3+}	is indicated as Cr(III)	So, Cr_2O_3	is written as	Chromium(III) oxide
V^{5+}	is indicated as V(V)	So, V_2O_5	is written as	Vanadium(V) oxide
Fe^{2+}	is indicated as Fe(II)	So, $FeSO_4$	is written as	Iron(II) sulphate
Fe^{3+}	is indicated as Fe(III)	So, $Fe_2(SO_4)_3$	is written as	Iron(III) sulphate
Mn^{7+}	is indicated as Mn(VII)	So, Mn_2O_7	is written as	Manganese(VII) oxide
Sn^{4+}	is indicated as Sn(IV)	So, SnO_2	is written as	Tin(IV) oxide
Sn^{2+}	is indicated as Sn(II)	So, SnO	is written as	Tin(II) oxide
Cr^{6+}	is indicated as Cr(VI)	So, $K_2Cr_2O_7$	is written as	Potassium dichromate(VI)
Cr^{6+}	is indicated as Cr(VI)	So, Na_2CrO_4	is written as	Sodium chromate(VI)

Self Assessment 1

1. In which reaction electrons are gained by the chemical species?
2. In which process hydrogen or any other electropositive radical is added?
3. What is the sign of oxidation number when electrons are lost or shifted partially away from an atom?
4. In which type of compounds, the oxidation number of hydrogen is –1?
5. How is Cu^{2+} described in the Stock notation?

BALANCING OF REDOX EQUATIONS

In a redox reaction, oxidation and reduction occur simultaneously.

- Oxidation involves an increase in the oxidation number.
- Reduction involves a decrease in the oxidation number.

Electrical charges are conserved during a redox reaction. Therefore, **in a redox reaction, the total gain in the oxidation numbers must be equal to the total loss in the oxidation numbers.**

Redox equations can be conveniently balanced by the following methods:

- Oxidation number method
- Ion-electron method

OXIDATION NUMBER METHOD

Oxidation number method of balancing chemical equations is based on the principle of conservation of **electrical charge** and **mass** during the course of a reaction.

Oxidation number method of balancing a chemical equation involves the following steps:

Step 1. Write the skeleton equation for the redox reaction.

Step 2. Write the oxidation number of each element in all the compounds above or below the symbol of the element in the formulae or symbols.

Step 3. Identify the elements which have undergone change in the oxidation number.

Usually, only two elements undergo change in their oxidation numbers — one gains in the oxidation number, while the other loses.

Step 4. Calculate the increase or decrease in the oxidation number per atom of the elements involved in the reaction.

Step 5. Multiply the change in the oxidation number by the number of atoms of the element involved in the reaction.

Step 6. Balance the total gain and loss in the oxidation numbers.

This is done by placing the number describing change in the oxidation number of oxidising agent in front of formula of reducing agent and vice versa.

Step 7. Complete the balancing of the two elements (which have changed oxidation number) on the right-hand side of the equation, maintaining the ratio of the oxidising agent to the reducing agent.

Step 8. Balance the oxygen atoms on both sides by adding H_2O to the required side.

Step 9. Balance the hydrogen atoms by adding H^+ ions to the required side.

The following examples illustrate the various steps involved in the balancing of chemical equation for a redox reaction by oxidation number method.

Reaction between $AgNO_3(aq)$ and Copper Metal

Step 1. The skeleton equation for this reaction is,

$$Cu(s) + AgNO_3(aq) \rightarrow Cu(NO_3)_2(aq) + Ag(s)$$

Step 2. Writing the oxidation numbers of the main species involved in the reaction,

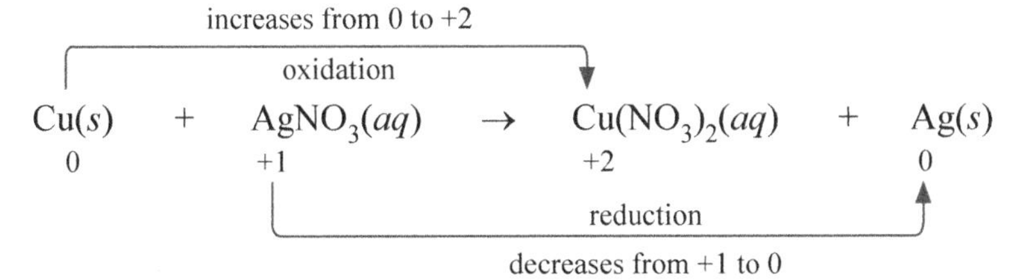

NO_3^- ion remains unchanged during the reaction.

Step 3. From the above equation,

Increase in the oxidation number of Cu atom = 2 – 0 = 2

Decrease in the oxidation number of Ag atom = 1 – 0 = 1

Step 4. To balance the total change in the oxidation numbers, multiply $AgNO_3(aq)$ by 2. Then, one can write,

$$Cu(s) + 2AgNO_3(aq) \rightarrow Cu(NO_3)_2(aq) + Ag(s)$$

Step 5. To balance Ag on both sides, multiply Ag on the right-hand side by 2. Then,

$$Cu(s) + 2AgNO_3(aq) \rightarrow Cu(NO_3)_2(aq) + 2Ag(s)$$

This is the balanced chemical equation for the given redox reaction.

Reaction between Na_2SO_3 and $KMnO_4$ in the presence of H_2SO_4

Step 1. The skeleton equation for this reaction is,

$$Na_2SO_3 + KMnO_4 + H_2SO_4 \rightarrow K_2SO_4 + MnSO_4 + Na_2SO_4 + H_2O$$

Step 2. Writing the oxidation numbers of the main species involved in the reaction,

decreases from +7 to +2

reduction

$$\underset{+7}{KMnO_4} + H_2SO_4 + \underset{+4}{Na_2SO_3} \rightarrow K_2SO_4 + \underset{+2}{MnSO_4} + \underset{+6}{Na_2SO_4} + H_2O$$

oxidation

increases from +4 to +6

Step 3. The oxidation number of S atom increases from +4 (in Na_2SO_3) to +6 (in Na_2SO_4). The oxidation number of Mn atom decreases from +7 (in $KMnO_4$) to +2 (in $MnSO_4$). Therefore,

Increase in the oxidation number of S atom = 6 – 4 = 2

Decrease in the oxidation number of Mn atom = 7 – 2 = 5

Step 4. To balance the loss and gain in the oxidation numbers, $KMnO_4$ should be multiplied by 2 and Na_2SO_3 by 5. Then, one can write,

$$2KMnO_4 + H_2SO_4 + 5Na_2SO_3 \rightarrow K_2SO_4 + MnSO_4 + Na_2SO_4 + H_2O$$

Step 5. All other elements are balanced as follows:

- Mn on the right-hand side is balanced by multiplying $MnSO_4$ by 2,
- Na^+ is balanced by multiplying Na_2SO_4 by 5,
- SO_4^{2-} ion is balanced by multiplying H_2SO_4 by 3, and
- Hydrogen atoms are balanced by multiplying H_2O by 3.

Step 6. The resulting equation is the balanced equation for the given redox reaction.

$$2KMnO_4 + 3H_2SO_4 + 5Na_2SO_3 \rightarrow K_2SO_4 + 2MnSO_4 + 5Na_2SO_4 + 3H_2O$$

Reaction between $FeSO_4$ and $K_2Cr_2O_7$ in the presence of H_2SO_4

Step 1. The skeleton equation for this reaction is,

$$FeSO_4 + K_2Cr_2O_7 + H_2SO_4 \rightarrow Fe_2(SO_4)_3 + Cr_2(SO_4)_3 + K_2SO_4 + H_2O$$

Step 2. Writing the oxidation numbers of the main species involved in the reaction,

decreases from +6 to +3

reduction

$$\underset{+2}{FeSO_4} + \underset{+6}{K_2Cr_2O_7} + H_2SO_4 \rightarrow \underset{+3}{Fe_2(SO_4)_3} + \underset{+3}{Cr_2(SO_4)_3} + K_2SO_4 + H_2O$$

oxidation

increases from +2 to +3

Step 3. In this reaction,

- The oxidation number of Fe increases from +2 (in $FeSO_4$) to + 3 [in $Fe_2(SO_4)_3$].
- The oxidation number of Cr decreases from +6 (in $K_2Cr_2O_7$) to +3 [in $Cr_2(SO_4)_3$].

So,

Increase in the oxidation number of Fe per atom $= 3 - 2 = 1$

Decrease in the oxidation number of Cr per atom $= 6 - 3 = 3$

There are **two atoms** of **Cr** in one molecule of $K_2Cr_2O_7$. So,

Total decrease in the oxidation number of Cr atoms $= 2 \times 3 = 6$

Thus, the changes in the oxidation states are in the ratio 1 : 6 for Fe : Cr.

Step 4. To balance the total loss and gain of electrons, multiply $FeSO_4$ by 6, and $K_2Cr_2O_7$ by 1. Then, one can write,

$$6\,FeSO_4 + K_2Cr_2O_7 + H_2SO_4 \rightarrow Fe_2(SO_4)_3 + Cr_2(SO_4)_3 + K_2SO_4 + H_2O$$

Step 5. All other elements can be balanced as follows:

- Fe on RHS by multiplying $Fe_2(SO_4)_3$ by 3,
- SO_4^{2-} on LHS by multiplying H_2SO_4 by 7, and
- H and O atoms by multiplying H_2O on RHS by 7.

Thus, the balanced equation for the given redox reaction is,

$$6\,FeSO_4 + K_2Cr_2O_7 + 7\,H_2SO_4 \rightarrow 3Fe_2(SO_4)_3 + Cr_2(SO_4)_3 + K_2SO_4 + 7\,H_2O$$

BALANCING BY ION-ELECTRON METHOD

Balancing of a chemical equation by **ion-electron** method (or **half-reaction** method) can be done by following steps described below:

Step 1. Write down the oxidation numbers of the elements which appear in the given skeleton equation.

Step 2. Identify the species getting oxidised and that getting reduced from the oxidation numbers of the element(s) which occur in them.

Step 3. Split the whole equation into two half-equations — one for the oxidation reaction, while the other for the reduction reaction.

Step 4. Balance both the half-equations separately as follows:

a. Balance all atoms other than H and O.

b. Add electrons to either side to balance the ionic charges.

c. Balance H and O by adding H^+ or H_2O on the side, as required.

Step 5. Multiply one or both the half-equations with suitable factor so that the electrons are balanced.

Step 6. Add the two balanced and matched half-equations to get the balanced equation for the overall reaction.

OXIDATION-REDUCTION (REDOX) REACTIONS IN ACIDIC, BASIC OR NEUTRAL SOLUTIONS

The method of balancing redox reactions in acidic and basic solutions is slightly different. In such cases, the following rules can be followed:

- If H^+ or any acid appears on either side of the equation, the reaction takes place in the acidic solution.
- If OH^-, or any base appears on either side of the equation, the reaction takes place in the basic solution.
- If neither H^+, OH^- nor any acid or base is present in the solution, the reaction takes place in a neutral solution.

The method of balancing such redox reactions is illustrated in the examples given below:

Example 1: Write the skeleton equation for a reaction between sulphite ion (SO_3^{2-}) with permanganate ion (MnO_4^-) and balance it by ion-electron method.

reduction (MnO_4^- → Mn^{2+})

$$SO_3^{2-} + MnO_4^- \rightarrow Mn^{2+} + SO_4^{2-}$$

	SO_3^{2-}	MnO_4^-	Mn^{2+}	SO_4^{2-}
Name of the ion:	sulphite	permanganate	manganous	sulphate
O.N. of the central atom:	+4	+7	+2	+6

oxidation (SO_3^{2-} → SO_4^{2-})

Here, SO_3^{2-} ion is getting oxidised to SO_4^{2-}, while MnO_4^- is getting **reduced** on Mn^{2+}.

The balancing of this equation by **ion-electron method** is illustrated on the next page.

Step	Oxidation	Reduction
1. Identify the two half-reactions	$SO_3^{2-} \rightarrow SO_4^{2-}$	$MnO_4^- \rightarrow Mn^{2+}$
2. Balancing all atoms other than H and O	S is already balanced.	Mn is already balanced.
3. Balancing electrons	$SO_3^{2-} \rightarrow SO_4^{2-} + 2e^-$	$MnO_4^- + 5e^- \rightarrow Mn^{2+}$
4. Balancing electrical charge by adding H^+	$SO_3^{2-} \rightarrow SO_4^{2-} + 2H^+ + 2e^-$	$MnO_4^- + 8H^+ + 5e^- \rightarrow Mn^{2+}$
5. Balancing O and H by adding H_2O	$SO_3^{2-} + H_2O \rightarrow SO_4^{2-} + 2H^+ + 2e^-$	$MnO_4^- + 8H^+ + 5e^- \rightarrow Mn^{2+} + 4H_2O$
6. Check, if the two half-equations are balanced or not	This half-equation is balanced.	This half-equation is balanced.
7. Balancing the number of electrons in the two half-equations	Multiply this half-equation by 5. $5SO_3^{2-} + 5H_2O \rightarrow 5SO_4^{2-} + 10H^+ + 10e^-$	Multiply this half-equation by 2. $2MnO_4^- + 16H^+ + 10e^- \rightarrow 2Mn^{2+} + 8H_2O$
8. Adding the two balanced half-equations and cancelling the common terms	$\mathbf{5SO_3^{2-} + 2MnO_4^- + 6H^+ \rightarrow 5SO_4^{2-} + 2Mn^{2+} + 3H_2O}$ This is the **balanced redox equation.**	

Example 2: Balance the following redox reaction:

$$Cu + NO_3^- \rightarrow NO_2 + Cu^{2+}$$

The given redox reaction can be written as follows:

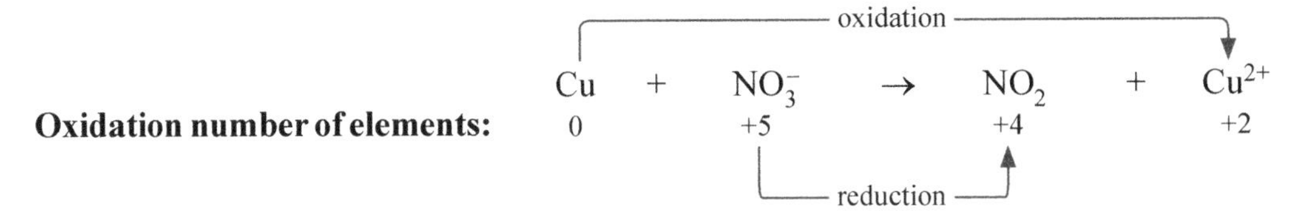

Here, Cu gets oxidised to Cu^{2+}

and NO_3^- ion gets reduced to NO_2.

Balancing of this equation by ion-electron method is illustrated on the next page.

Step	Oxidation half-reaction	Reduction half-reaction
1. Identifying half-reactions	$Cu \rightarrow Cu^{2+}$	$NO_3^- \rightarrow NO_2$
2. Balancing all atoms other than H and O	Cu is balanced.	N is balanced.
3. Balancing electrons (oxidation numbers are given below each species)	$Cu \rightarrow Cu^{2+} + 2e^-$ 0 +2	$NO_3^- + e^- \rightarrow NO_2$ +5 +4
4. Balancing electrical charges by adding H^+	Charges are balanced	$NO_3^- + 2H^+ + e^- \rightarrow NO_2$
5. Balancing O and H by adding H_2O	–	$NO_3^- + 2H^+ + e^- \rightarrow NO_2 + H_2O$
6. Check, if the two half-equations are balanced	This half-equation is balanced.	This half-equation is balanced.
7. No. of electrons involved	2	1
8. Balancing the number of electrons in the two half-equations	– $Cu \rightarrow Cu^{2+} + 2e^-$	Multiply this half-equation by 2. $2NO_3^- + 4H^+ + 2e^- \rightarrow 2NO_2 + 2H_2O$
9. Adding the two balanced half-equations and cancelling the common terms	$\mathbf{Cu + 2NO_3^- + 4H^+ \rightarrow Cu^{2+} + 2NO_2 + 2H_2O}$ This is the **balanced redox equation.**	

Example 3: Balance the chemical equation for the following redox reaction by ion-electron method.

$$Cr(OH)_4^- + H_2O_2 \rightarrow CrO_4^{2-} + H_2O \quad \text{(basic solution)}$$

The given redox reaction can be described as follows:

oxidation: $Cr(OH)_4^- \rightarrow CrO_4^{2-}$

$$Cr(OH)_4^- + H_2O_2 \rightarrow CrO_4^{2-} + H_2O \quad \text{(in basic solution)}$$

Oxidation number of elements: +3 (Cr in $Cr(OH)_4^-$), −1 (O in H_2O_2), +6 (Cr in CrO_4^{2-}), −2 (O in H_2O)

reduction: $H_2O_2 \rightarrow H_2O$

Here, Cr^{3+} (in $Cr(OH)_4^-$) gets oxidised to Cr^{6+} (in CrO_4^{2-}).

and O_2^{2-} (in H_2O_2) gets reduced to O^{2-} (in H_2O).

Step	Oxidation half-reaction	Reduction half-reaction
1. Identifying half-reactions	$Cr(OH)_4^- \rightarrow CrO_4^{2-}$	$H_2O_2 \rightarrow H_2O$
2. Balancing all atoms other than H and O	Cr is balanced.	–
3. Balancing electrons in each half-equation*	$Cr(OH)_4^- \rightarrow CrO_4^{2-} + 3e^-$ (+3) (+6)	$H_2O_2 + 2e^- \rightarrow H_2O$ (–1) (–2)
4. Balancing electrical charges by adding OH^- (basic solution)	$Cr(OH)_4^- + 4OH^- \rightarrow CrO_4^{2-} + 3e^-$	$H_2O_2 + 2e^- \rightarrow H_2O + 2OH^-$
5. Balancing O and H by adding H_2O	$Cr(OH)_4^- + 4OH^- \rightarrow CrO_4^{2-} + 4H_2O + 3e^-$	$H_2O_2 + H_2O + 2e^- \rightarrow H_2O + 2OH^-$ or $H_2O_2 + 2e^- \rightarrow 2OH^-$
6. Check, if the two half-equations are balanced	This half-equation is balanced w.r.t all elements.	This half-equation is balanced w.r.t all elements.
7. No. of electrons involved	Three ($3e^-$)	Two ($2e^-$)
8. Balancing the number of electrons in the two half-equations	Multiply the above equation by 2 $2 \times [Cr(OH)_4^- + 4OH^- \rightarrow CrO_4^{2-} + 4H_2O + 3e^-]$	Multiplying the above equation by 3 $3 \times [H_2O_2 + 2e^- \rightarrow 2OH^-]$
9. Adding the two balanced half-equations and cancelling the common terms	**$2Cr(OH)_4^- + 2OH^- + 3H_2O_2 \rightarrow 2CrO_4^{2-} + 8H_2O$** This is the **balanced redox equation.**	

* The numbers written below the elements are their oxidation numbers.

Self Assessment 2

1. In which type of reactions, the total increase in oxidation numbers is equal to the total decrease in the oxidation numbers?
2. On which two parameters, is the balancing of chemical equations by oxidation number method based?
3. In which method of balancing chemical equations, the overall chemical equation is split into two half-equations?
4. Name the two half-equations generated from an overall chemical equation.
5. Rewrite the chemical equation:

$$CuSO_4(aq) + H_2S(aq) \rightarrow CuS(s) + H_2SO_4(aq)$$

and balance it by oxidation number method.

6. Balance the redox reaction:

(a) $SnO_2(s) + C(s) \xrightarrow{\Delta} Sn(l) + CO(g)$

(b) $Fe_2O_3(s) + CO(g) \xrightarrow{\Delta} Fe(l) + CO_2(g)$ by ion-electron method.

REVIEW QUESTIONS

A. Very Short Answer Type Questions

1. Give one example of an oxidising agent.
2. What is the most common oxidation state of calcium?
3. Why is the following reaction considered reduction of CuO?

$$CuO(s) + H_2(g) \xrightarrow{\Delta} Cu(s) + H_2O(l)$$

4. What is the oxidation number of copper in copper sulphate?
5. What is the oxidation number of oxygen in peroxides?
6. In which form all the elements have zero oxidation number?
7. Describe sodium chromate in terms of Stock notation.
8. Split the reaction:

$$SO_3^{2-}(aq) + MnO_4^-(aq) \rightarrow Mn^{2+}(aq) + SO_4^{2-}(aq)$$

into two half-equations.

9. How would you identify the species getting oxidised?

B. Short Answer Type Questions

1. What is meant by a redox reaction? Give one example.
2. What is a reducing agent? Give one example.
3. Define oxidation number. When is the oxidation number of an element positive?
4. Explain why a substance that gains electrons is said to be reduced.
5. What is the half-equation method of balancing chemical equations?

C. Long Answer Type Questions

1. What is the oxidation number of S in:

   (a) K_2SO_3 (b) K_2SO_4 (c) $KHSO_4$?

2. Define oxidation and reduction in terms of electronic concept.
3. State the rules for counting of electrons when an atom is removed from any chemical species.
4. Compare the oxidation numbers of anions in HF, HClO, $HBrO_2$ and HIO_4.
5. How is the oxidation number of a metal indicated in terms of Stock notation?
6. Illustrate the balancing of a chemical equation by ion-electron method.
7. Split the chemical equation given below into two half-equations.

$$Cu + NO_3^- \rightarrow NO_2 + Cu^{2+}$$

Identify the oxidation half-equation and the reduction half-equation.

D. True or False Type Questions

Write 'T' for True or 'F' for False.

1. Redox reactions do not involve any subatomic particle.
2. Oxidation number of fluorine in the combined state is –1.
3. When an atom loses electrons, its oxidation number increases.
4. Fluorine is the strongest reducing agent.
5. A chemical change involving loss of electrons is known as reduction.

6. A good reducing agent must be easily reducible.
7. Oxidation number of hydrogen in a metal hydride is –1.
8. Oxidation number of an element in its monatomic ion is equal to the charge on the ion.
9. Ion-electron method of balancing chemical equations is also known as the half-reaction method.
10. The half-reaction in which the reactant loses one or more electrons is called oxidation half-reaction.

E. Fill in the Blanks Type Questions

1. Oxidation number of elements in the uncombined state is always ____________
2. In a polyatomic ion, the sum of the oxidation numbers of all the atoms is equal to ________ on it.
3. When an atom accepts electrons, the oxidation number will ______________
4. A chemical change involving ________ of electrons is known as oxidation.
5. Write the most common oxidation state of the elements / ions given below.

Element/ion	Oxidation state	Element/ion	Oxidation state
H	....	He	....
C	....	B	....
P	....	N	....
O	....	F	....
Li	....	Mg	....
Ca^{2+}	....	Na	....
K	....	Be	....
Al	....	Cd	....
Ag	....	NH_4^+	....
SO_4^{2-}	....	PO_4^{3-}	....
Cl	....	CO_3^{2-}	....
Hg^{2+}	....	BO_3^{3-}	....

F. Match the Columns Type Questions

a.

	Column A		Column B
1	Addition of oxygen	A.	Oxidising agent
2	Chlorine	B.	Zero oxidation number
3	Element	C.	Oxidation
4	Oxidation number of F	D.	Sn(IV)
5	Stannic	E.	–1

b.

	Column A		Column B
1	$K_2Cr_2O_7$	A.	Reduction
2	Addition of hydrogen	B.	Cr^{6+}
3	Positive oxidation number	C.	Reducing agent
4	Calcium	D.	Strongest oxidising agent
5	Fluorine	E.	Electropositive element

G. Multiple Choice Type Questions

Choose the correct alternative.

1. Which subatomic particle is involved in all redox reactions?
   (a) neutron ❑ (b) proton ❑ (c) electron ❑ (d) none of these ❑
2. An oxidising agent is
   (a) reduced as it gains electrons. ❑ (b) oxidised as it loses electrons. ❑
   (c) reduced as it loses electrons. ❑ (d) oxidised as it gains electrons. ❑
3. In a redox reaction, the species which loses electrons
   (a) is oxidised. ❑ (b) is reduced. ❑
   (c) gains mass at the electrode. ❑ (d) decreases in oxidation number. ❑
4. In the following reaction, the species being reduced is,
   $$2FeBr_3 + 3Cl_2 \rightarrow 2FeCl_3 + 3Br_2$$
   (a) Cl in Cl_2 ❑ (b) Fe in $FeBr_3$ ❑ (c) Br in $FeBr_3$ ❑ (d) none of these ❑
5. The oxidation number of oxygen in oxyfluorides is
   (a) 0 ❑ (b) −1 ❑ (c) −2 ❑ (d) +2 ❑
6. The chemical species which can give electrons to some other species is called
   (a) oxidising agent ❑ (b) reducing agent ❑
   (c) electronating agent ❑ (d) none of these ❑
7. In which compound is nitrogen in the +1 oxidation state?
   (a) N_2O_4 ❑ (b) N_2 ❑ (c) N_2O ❑ (d) none of these ❑
8. In the reaction, $2Na(s) + H_2(g) \rightarrow 2NaH(s)$, the oxidising agent is
   (a) $Na(s)$ ❑ (b) $NaH(s)$ ❑ (c) $H_2(g)$ ❑ (d) none of these ❑
9. Which of the following substances is oxidised during the following reaction?
   $$Fe + H_2SO_4 \rightarrow FeSO_4 + H_2$$
   (a) Fe ❑ (b) H_2SO_4 ❑ (c) $FeSO_4$ ❑ (d) H_2 ❑
10. Consider the following redox reaction:
    $$As_2O_3 + 2NO_3^- + 2H_2O + 2H^+ \rightarrow 2H_3AsO_4 + N_2O_3$$
    In this reaction, nitrogen of NO_3^-
    (a) loses electrons and increases oxidation number. ❑
    (b) gains electrons and increases oxidation number. ❑
    (c) loses electrons and decreases oxidation number. ❑
    (d) gains electrons and decreases oxidation number. ❑

ANSWERS

A. 1. Oxygen 2. +2 3. Oxygen is removed from CuO.
4. +2 5. −1 6. Elemental form
7. Sodium chromate (VI) 8. $SO_3^{2-} \rightarrow SO_4^{2-} + 2e^-$; $MnO_4^- + 5e^- \rightarrow Mn^{2+}$
9. Increase in oxidation number

D. 1. F 2. T 3. T 4. F 5. F 6. F 7. T 8. T 9. T 10. T

E. 1. Zero 2. Charge 3. Decrease 4. Loss of electrons
5. +1, +4, +3, +5, −2, +1, +2, +1, +3, +1, −2, −1, +2, 0, +3, +3, +5, −1, +2, +1, +2, +2, +1, −3, −2, −3

F. a. 1 – C; 2 – A; 3 – B; 4 – E; 5 – D
b. 1 – B; 2 – A; 3 – E; 4 – C; 5 – D

G. 1. c 2. a 3. a 4. a 5. d 6. b 7. c 8. c 9. a 10. d

Conical (Chemistry) Crossword

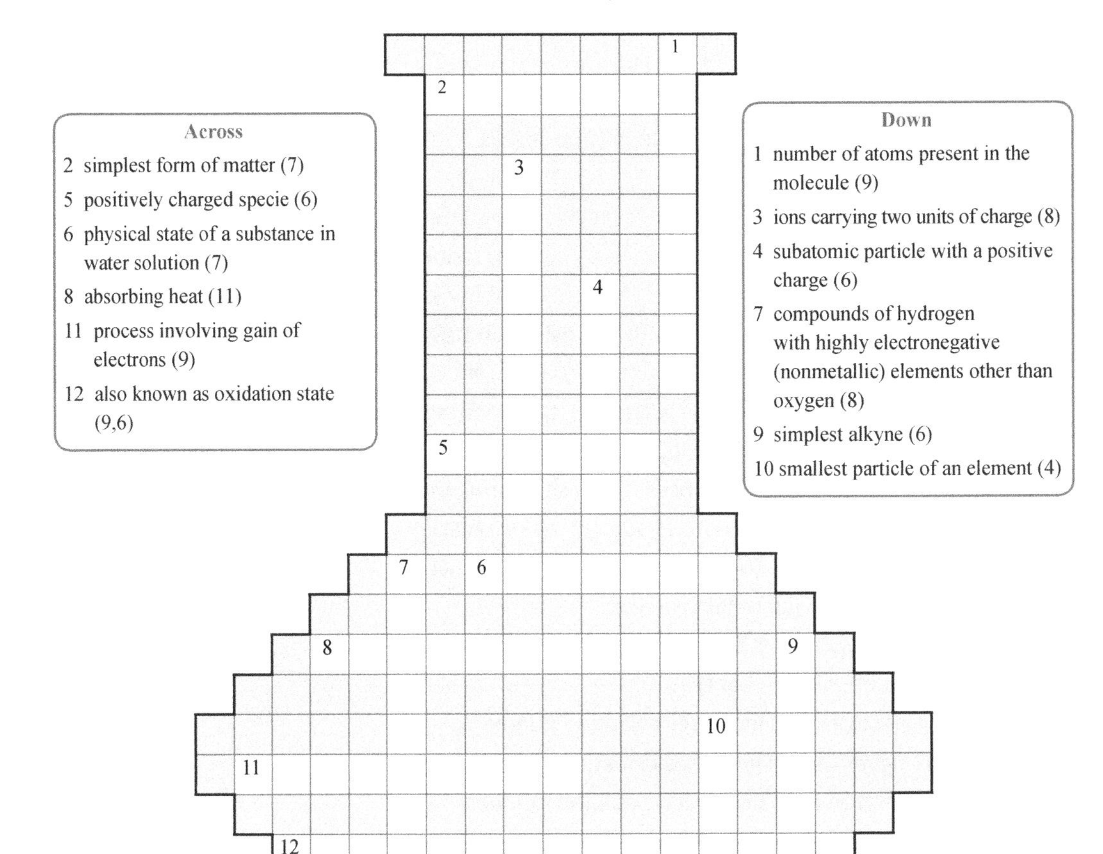

10

Nomenclature for Organic Compounds

ORGANIC CHEMISTRY — The branch of Chemistry which deals with the study of compounds of carbon with hydrogen and their derivatives is called **Organic Chemistry.**

ORGANIC COMPOUNDS — The compounds obtained from animals or plants directly or indirectly are called **organic compounds.**

There are about five million (fifty lakh) compounds of carbon that are known to us.

FUNCTIONAL GROUP — An atom or a group of atoms which gives characteristic properties to a compound is called a **functional group.**

HOMOLOGOUS SERIES — A group of organic compounds showing similar chemical properties, can be described by a common general formula or/and has a particular functional group is called a **homologous series.** Each group/family of organic compounds is characterised by a **functional group** present in it. The functional groups corresponding to various families (homologous series) of organic compounds are given below:

Some Simple Functional Groups

Homologous series	Characteristic functional group	
	Name	Formula
1. Hydrocarbon		
❖ alkane	Single bond	$C-C$
❖ alkene	Double bond	$>C=C<$
❖ alkyne	Triple bond	$-C \equiv C-$
2. Halogen derivatives of hydrocarbons	Halo	$-X$ (X = Cl, Br or I)
3. Alcohol	Hydroxyl	$-OH$
4. Aldehyde	Aldehydic	$-CHO$
5. Ketone	Ketonic	$>C=O$
6. Carboxylic acid	Carboxyl	$-COOH$
7. Ether	Ether	$\Rightarrow C-O-C \Leftarrow$
8. Ester	Ester	$-CO-OR$

HYDROCARBONS – The compounds consisting of only **carbon and hydrogen** are called **hydrocarbons.** Hydrocarbons can be classified into the following classes:

- Saturated hydrocarbons
- Unsaturated hydrocarbons
- Straight-chain hydrocarbons
- Branched-chain hydrocarbons

SATURATED HYDROCARBONS — Hydrocarbons in which all carbon atoms are bonded to each other by single covalent bonds are called **saturated hydrocarbons.**

Saturated hydrocarbons are called alkanes.

Alkanes are described by the general formula C_nH_{2n+2}.

Some simple saturated hydrocarbons are:

- Methane (CH_4)
- Ethane (C_2H_6)
- Propane (C_3H_8)
- Butane (C_4H_{10})

UNSATURATED HYDROCARBONS – Hydrocarbons in which two carbon atoms are bonded to each other by a double bond (=) or a triple bond (≡) and carbon atoms are bonded to hydrogen atoms by single covalent bonds are called **unsaturated hydrocarbons.**

Unsaturated hydrocarbons are of two types:

- Alkenes
- Alkynes

ALKENE – Hydrocarbons in which two carbon atoms are bonded to each other by a **double** bond (=) are called alkenes.

Thus, an alkene contains a $>C = C<$ group.

ALKYNE – Hydrocarbons in which two carbon atoms are bonded to each other by a **triple bond** (≡) are called alkynes.

Thus, an alkyne contains a $-C \equiv C-$ group.

Typical unsaturated hydrocarbons are:

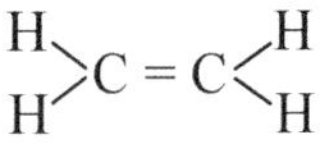

$$\begin{matrix} H \\ H \end{matrix}\!>\!C = C\!<\!\begin{matrix} H \\ H \end{matrix} \quad \text{or} \quad H_2C = CH_2$$

ethene (ethylene)
(It contains a carbon-carbon double (=) bond.)

$$H - C \equiv C - H \quad \text{or} \quad HC \equiv CH$$

ethyne (acetylene)
(It contains a carbon-carbon triple (≡) bond.)

STRAIGHT-CHAIN HYDROCARBONS – Hydrocarbons in which all carbon atoms are bonded to each other along a continuous straight chain of carbon atoms are called **straight-chain hydrocarbons.**

The hydrocarbon described by the formula

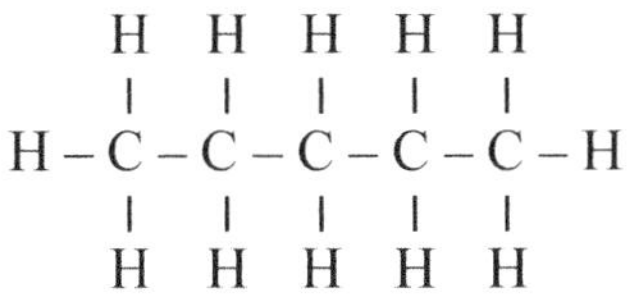

$$\begin{matrix} & H & & H & & H & & H & & H & \\ & | & & | & & | & & | & & | & \\ H - & C & - & C & - & C & - & C & - & C & - H \\ & | & & | & & | & & | & & | & \\ & H & & H & & H & & H & & H & \end{matrix}$$

is a straight-chain hydrocarbon.

BRANCHED-CHAIN HYDROCARBONS – Hydrocarbons in which one or more H atoms in the straight chain of carbon atoms are replaced by one or more alkyl group (e.g., $-CH_3$, $-C_2H_5$, etc.) are called **branched-chain hydrocarbons.**

The hydrocarbon described below has a branching in its chain (shown in colour) and it is a branched-chain hydrocarbon.

```
                H   Branch
                |  /
            H - C - H
    H   H   |   H
    |   |   |   |
H - C - C - C - C - H     Main Chain
    |   |   |   |
    H   H   H   H
```

NOMENCLATURE OF HYDROCARBONS – Naming of compounds on the basis of conventions or certain rules is called **nomenclature** of compounds. The International Union of Pure and Applied Chemistry (IUPAC) has devised a systematic method of naming organic compounds. The rules for naming organic compounds based on the **IUPAC Recommendations 1993** are described here.

NAMING STRAIGHT-CHAIN HYDROCARBONS

To name a straight-chain hydrocarbon, we need to know the following:

- The number of carbon atoms present in the longest continuous chain of carbon.
- The nature of hydrocarbon, i.e., whether it is saturated or unsaturated.

> The number of carbon atoms in the molecule of a hydrocarbon is indicated by its **stem or word root.**
>
> The stem (or word root) is the starting part of the hydrocarbon name.

To name a straight-chain hydrocarbon, follow the steps described below:

Step 1. Count the number of carbon atoms in the longest continuous chain.

Step 2. Determine the stem (or the word root) for the molecule by using the information given below.

No. of carbon atoms :	1	2	3	4	5	6	7	8
Stem (or word root):	Meth	Eth	Prop	But	Pent	Hex	Hept	Oct

Step 3.
- The nature of hydrocarbon is described by a **suffix.** The use of suffix for various types of hydrocarbons is illustrated below:
- A saturated hydrocarbon or an alkane is indicated by the suffix **-ane** which is added at the **end** of the **stem** (or the **word root).**

> **Prefix** is the term which is added in the **beginning of the name.**
>
> **Suffix** is the term which is added at the **end of the name.**

- The compound CH_4, which contains only **one carbon atom,** is named as follows:

IUPAC name of CH_4 ⇒ Word root + Suffix ⇒ Meth + ane ⇒ Methane

- The compound C_2H_6, which contains **two carbon atoms,** is named as follows:

IUPAC name of C_2H_6 ⇒ Word root + ane ⇒ Eth + ane ⇒ Ethane

- An unsaturated hydrocarbon containing a double bond (=), that is, an **alkene** is indicated by the suffix –**ene**. The suffix –**ene** is added at the **end** of the **stem (or word root).**

- The compound $CH_2 = CH_2$ (or C_2H_4), which contains **two carbon atoms,** is named as follows:

IUPAC name of C_2H_4 ⇒ Word root + Suffix ⇒ Eth + ene ⇒ Ethene

Similarly, the compound C_3H_6, which contains **three carbon atoms,** is named as follows:

IUPAC name of C_3H_6 ⇒ Word root + Suffix ⇒ Prop + ene ⇒ Propene

- An unsaturated hydrocarbon containing a triple bond (≡), that is, an **alkyne** is indicated by the suffix **-yne.** The suffix **-yne** is added at the **end** of the **stem** (or **word root).**

- The compound H – C = C – H (or C_2H_2), which contains **two carbon atoms,** is named as follows:

IUPAC name of C_2H_2 ⇒ Word root + Suffix ⇒ Eth + yne ⇒ Ethyne

Similarly, the compound C_3H_4, which contains **three carbon atoms,** is named as follows:

IUPAC name of C_3H_4 ⇒ Word root + Suffix ⇒ Prop + yne ⇒ Propyne

NAMING THE BRANCHED-CHAIN HYDROCARBONS

Naming a branched-chain hydrocarbon is done as follows:

Step 1. ❖ **The longest chain rule** - The longest continuous chain of carbon atoms in the structure of the molecule is called the **longest chain.**

- Hydrocarbon which corresponds to the longest carbon chain is called the **parent hydrocarbon.**
- The given compound is then named as a derivative of the parent hydrocarbon.

Step 2. ❖ **The lowest number rule** - The alkyl group or groups present in the side chain are considered as the **substituents.**

- The carbon atoms of the longest carbon chain are numbered in such a way that the carbon atom bonded to the **substituent** gets the **lowest** possible number.

Step 3. ❖ **Location of the substituent** - Position of the substituent is indicated by the serial **number** (called **locant)** of the carbon atom to which it is attached.

Step 4. ❖ **Naming the compound** - The IUPAC name of the compound is then obtained by writing in the following order:

Name of the branched chain hydrocarbon = Locant-prefix of the substituent + Name of the parent hydrocarbon

This is illustrated by the following example:

Name the following compound:

$$H_3C - \underset{\displaystyle CH_3}{\underset{|}{CH}} - CH_3$$

❖ Inspection of the structure indicates that the longest carbon chain in this structure contains **three carbon atoms.** Therefore, the **parent hydrocarbon** for this compound is **propane.**

❖ In this molecule, one **methyl group** ($-CH_3$) is present as the side chain. Therefore, methyl group is the substituent.

❖ Numbering of carbon atoms gives four choices as follows:

$$H_3{}^3C - \overset{\displaystyle H}{\overset{|}{\underset{\displaystyle CH_3}{\underset{|}{{}^2C}}}} - {}^1CH_3 \qquad H_3C - \overset{\displaystyle H}{\overset{|}{\underset{\displaystyle {}^3CH_3}{\underset{|}{{}^2C}}}} - {}^1CH_3 \qquad H_3{}^1C - \overset{\displaystyle H}{\overset{|}{\underset{\displaystyle CH_3}{\underset{|}{{}^2C}}}} - {}^3CH_3 \qquad H_3C - \overset{\displaystyle H}{\overset{|}{\underset{\displaystyle {}^1CH_3}{\underset{|}{{}^2C}}}} - {}^3CH_3$$

In all the four choices, the $- CH_3$ group (the substituent) is at carbon atom number 2. Therefore, the lowest number rule gives position of the substituent (methyl group) as carbon atom number 2. So, the locant of the substituent is 2.

❖ IUPAC name of the given compound is written as follows:

Name of the compound ⇒ Locant-prefix of the substituent + Name of the parent hydrocarbon

Therefore,

Name of the compound ⇒ 2-methyl + propane ⇒ 2-methylpropane

Identifying a Hydrocarbon from its Name

To identify a hydrocarbon from its name, one should look at the **ending letters** in its name.

❖ If the name ends with **ane,** then it should be an **alkane** (saturated hydrocarbon containing carbon-carbon **single bonds).**

❖ If the name ends with **ene,** then it should be an **alkene** (unsaturated hydrocarbon containing at least one carbon-carbon **double bond).**

- If the name ends with **yne,** then it should be an **alkyne** (unsaturated hydrocarbon containing at least one carbon-carbon **triple bond**).

Identifying a Hydrocarbon from its Formula

To identify a hydrocarbon from its formula, one should look at a given formula of the hydrocarbon and find out:

- the number of carbon atoms, and
- the number of hydrogen atoms.

Then, proceed as follows:

Step 1. Count the number of C atoms in the hydrocarbon molecule.

Step 2. Count the number of H atoms in the hydrocarbon molecule.

Step 3. Then, use the information given below to identify the type of hydrocarbon.

The hydrocarbon identified by its formula is given in the following Table.

Nature and Type of Hydrocarbon

Formula of hydrocarbon	C_nH_{2n+2}	C_nH_{2n}	C_nH_{2n-2}
No. of C atoms	n	n	n
No. of H atoms	$2n+2$	$2n$	$2n-2$
Nature of hydrocarbon	Alkane	Alkene	Alkyne
Type of hydrocarbon	Saturated	Unsaturated	Unsaturated

NAMING THE COMPOUND HAVING A SUBSTITUENT (OTHER THAN AN ALKYL GROUP)

A compound having a substituent (other than an alkyl group) is named as follows:

Step 1. Identify the **longest continuous chain** of carbon atoms. This gives the name of the **parent hydrocarbon.**

Step 2. Choose the prefix for the substituent from Table given below:

Some common prefixes and suffixes are given below in Table.

The Prefixes and Suffixes of Some Substituents

Substituent	Prefix	Functional group	Suffix
Chlorine	Chloro	Alcoholic (–OH)	-ol
Bromine	Bromo	Aldehydic (–CHO)	-al
Iodine	Iodo	Ketonic ($>C=O$)	-one
		Carboxylic (–COOH)	-oic acid

Step 3. Number the carbon chain such that the carbon bonded to the substituent gets the lowest number. This numeral is called the **locant** of the substituent.

Step 4. Write name of the compound as follows:

Name of the compound ⇒ Locant-Prefix of the substituent + Parent hydrocarbon

This method of naming substituted hydrocarbon is illustrated below:

Example: Name the following compound:

$$\begin{array}{c} CH_3 - CH - CH_3 \\ | \\ I \end{array}$$

Follow the following steps:

Step 1. The given molecule has the longest chain of **three** carbon atoms. Therefore, the parent hydrocarbon is **propane.**

Step 2. The molecule has an atom of **iodine** bonded to carbon atom numbering 2. So, the locant of iodine is 2.

Step 3. The IUPAC name of the given compound is written as follows:

Name of the compound ⇒ Locant-Prefix of iodine + Name of the parent hydrocarbon

Therefore,

Name of the compound ⇒ 2-iodo + propane ⇒ 2-iodopropane

Self Assessment 1

1. Name the elements present in a hydrocarbon.
2. What are the general formulae of the following:
   (a) Alkane (b) Alkene (c) Alkyne
3. Illustrate by giving appropriate formulae, the
   (a) Straight chain hydrocarbon. (b) Branched chain hydrocarbon.
4. How will you identify the nature of a hydrocarbon from its name.
   (a) the name ends with **-ane** (b) the name ends with **-yne**

NAMING THE COMPOUND HAVING A FUNCTIONAL GROUP

A compound having a functional group is named as follows:

Step 1. Select the longest continuous chain of carbon atoms containing the **functional group.**

Step 2. Identify the parent hydrocarbon.

Step 3. Select the **suffix** for the functional group.

Step 4. Locate position of the functional group by numbering the C-chain such that the carbon atom having the functional group gets the lowest number. This number is called **locant**.

Step 5. Write IUPAC name of the compound as follows:

Name of the compound ⇒ Name of the parent hydrocarbon – e + Locant + Suffix

Naming Alcohols

General formula: ROH or $C_nH_{2n+1}OH$ **Functional group: -OH**

- Count the number of carbon atoms in the longest continuous chain containing the -OH group.
- From the number of carbon atoms in the longest chain, identify the parent alkane. Then,

IUPAC name of an alcohol = IUPAC name of the parent alkane – e + Locant + ol

Formula of alcohol	No. of carbon atoms in the longest chain	Parent hydrocarbon	Functional group present in alcohol	Suffix	IUPAC name of the compound
CH_3OH	1	Methane	-OH	-ol	Methane – e + ol = **Methanol**
C_2H_5OH	2	Ethane	-OH	-ol	Ethane – e + ol = **Ethanol**
$CH_3-CH_2-CH_2OH$	3	Propane	-OH	-ol	Propane – e + ol = **Propanol**
$CH_3-CH(OH)-CH_3$	3	Propane	-OH (at carbon atom 2)	-ol	Propane – e + -2-ol = **Propan-2-ol**

Naming Aldehydes

General formula: RCHO **Functional group: $-C(=O)H$**

- Count the number of carbon atoms in the longest continuous chain containing the -CHO group.
- Identify the parent alkane from the number of C-atoms in the chain.
- The chain is numbered by giving the carbon atom of the -CHO group number 1, if required. Then,

IUPAC name of the aldehyde = IUPAC name of the parent alkane – e + al = Alkanal

Formula of aldehyde	No. of carbon atoms in the longest chain	Parent alkane	Functional group present in the molecule	Suffix	IUPAC name of the aldehyde
HCHO	1	Methane	-CHO	-al	Methane – e + al = **Methanal**
CH_3CHO	2	Ethane	-CHO	-al	Ethane – e + al = **Ethanal**
C_2H_5CHO	3	Propane	-CHO	-al	Propane – e + al = **Propanal**

- In aldehydes, the **-CHO group** is always present at the **end of the chain.** So, there is no need to specify its position in the name.
- While counting the carbon atoms in the parent chain, the carbon of the -CHO group is also counted (as carbon number 1).

Naming Ketones

General formula: RCOR′ **Functional group: $>C=O$**

- Count the number of carbon atoms in the longest continuous chain containing the carbonyl ($>C=O$) group.
- Identify the parent alkane from the number of carbon atoms in the longest chain.
- The chain is numbered so that the carbonyl (aldehydic or ketonic) carbon gets the lowest number.

Then,

IUPAC name of the ketone = IUPAC name of the parent alkane − e + locant + one

Formula of ketone	No. of carbon atoms in the longest chain	Parent alkane	Functional group present in the molecule	Suffix	IUPAC name of the ketone
$^{1}CH_3 \cdot {}^{2}CO \cdot {}^{3}CH_3$	3	Propane	$>C=O$	-one	Propane – e + -2- + one = **Propan-2-one**
$^{1}CH_3 \cdot {}^{2}CO \cdot {}^{3}CH_2{}^{4}CH_3$	4	Butane	$>C=O$ (at carbon atom 2)	-one	Butane – e + -2- + one = **Butan-2-one**

Naming Carboxylic Acids

General formula: RCOOH **Functional group: $-C\begin{matrix}\nearrow O \\ \searrow OH\end{matrix}$**

- Select the longest continuous chain of carbon atoms containing the -COOH group.
- Count the number of C-atoms in the longest continuous chain of carbon atoms.
- The carbon chain is numbered by giving carbon atom of the -COOH group serial number 1.
- Identify name of the parent alkane from the number of carbon atoms in the longest continuous chain.

Then,

IUPAC name of the carboxylic acid = IUPAC name of the parent alkane – e + oic acid

Formula of carboxylic acid	No. of carbon atoms in the longest chain	Parent alkane	Functional group present in the molecule	Suffix	IUPAC name of the carboxylic acid
HCOOH	1	Methane	-COOH	-oic acid	Methane – e + oic acid = **Methanoic acid**
CH_3COOH	2	Ethane	-COOH	-oic acid	Ethane – e + oic acid = **Ethanoic acid**
CH_3CH_2COOH	3	Propane	-COOH	-oic acid	Propane – e + oic acid = **Propanoic acid**

Self Assessment 2

1. How is a functional group described in the name of a compound?
2. Write the suffixes for the following functional groups.
   -OH, -CHO, >C = O, -COOH
3. Write formulae of the functional groups present in the following compounds.
   (a) Alcohol (b) Carboxylic acid (c) Aldehyde (d) Ketone
4. Name the following compounds:
   (a) HCOOH (b) $CH_3–CH_2–CHO$
5. Identify the following for the molecule $CH_2Cl–CH_2–CH_2–OH$
   (a) Stem (or word root) (b) Prefix (c) Suffix

REVIEW QUESTIONS

A. Very Short Answer Type Questions

1. Write the names of the first three members of the alkene series.
2. Give name and molecular formula of the next higher homologue of propane.
3. Write formulae of the immediate higher and lower homologues of C_3H_4.
4. What is the IUPAC name of the following compound?

$$CH_3-\underset{\underset{Cl}{|}}{\overset{\overset{CH_3}{|}}{C}}-CH_3$$

5. What is the word root and suffix of an alkene containing three carbons?
6. Name the suffix used for saturated hydrocarbons.
7. Which type of hydrocarbons are characterised by the general formula C_nH_{2n}?
8. Name the first two members of the carboxylic acid family.

9. By what name an atom or a group of atoms which gives specific properties to a compound is called?
10. What is a group of organic compounds showing similar properties and a common general formula called?

B. Short Answer Type Questions

1. Write the names of the following organic compounds:
   (a) CH_2Cl_2
   (b) $CH_3CH(Cl)CH_3$
   (c) $CH_3CH_2C(CH_3)_2Br$
2. Write the molecular formula of the third and the fifth members of the homologous series of carbon compounds represented by C_nH_{2n-2}.
3. Name the functional groups present in the following compounds:
   (a) $CH_3CH_2CH_2OH$
   (b) $CH_3CH_2CH_2COOH$
   (c) CH_3CH_2CHO
   (d) $CH_3COCH_2CH_2CH_3$
   (e) $CH_3CH_2C(CH_3)_2I$
4. Write the formulae of the following organic molecules:
   (a) 3-ethylhexane (b) 2-ethylpent-2-ene
5. What is meant by a homologous series? Give an example.

C. Long Answer Type Questions

1. What is a functional group? Name two functional groups containing oxygen. Write their formulae also.
2. What are the saturated hydrocarbons? Write general formula for the saturated hydrocarbons.
3. How will you identify a hydrocarbon from its formula?
4. Name three compounds having the oxygen-containing functional groups. Write their formulae also.
5. (a) What are the branched-chain hydrocarbons? Describe the lowest number rule.
   (b) Which of the two is the correct way of numbering carbon atoms in a chain? Give reason.

```
            |                                  |
          - C -                              - C -
            |                                  |
(i) ⁴C - ³C - ²C - ¹C             (ii) ¹C - ²C - ³C - ⁴C
```

D. True or False Type Questions

Write 'T' for True or 'F' for False.

1. The IUPAC name of $CH_3CH_2CH_2COOH$ is propanoic acid.
2. Pentan-2-one and 2-pentanone are the same compounds.
3. The general formula for alkenes is C_nH_{2n-2}.
4. Methane, ethane, propane and butane are members of alkyne homologous series.
5. The organic compounds containing -COOH functional group are called esters.
6. The next higher homologue of butane is pentane.

7. The property of carbon atom by virtue of which it forms bonds with other carbon atoms is called polymerisation.
8. In alkanes, two carbon atoms are joined by an ionic bond.
9. Esters are the compounds containing carbon, hydrogen and oxygen.
10. Propane and propanone are homologues.

E. Fill in the Blanks Type Questions

1. The organic compound containing two carbons bonded by a double bond is called __________
2. The compounds consisting of only carbon and hydrogen are called __________
3. A member of any homologous series is called __________
4. The general formula of alkanes is __________
5. Saturated hydrocarbons are called __________
6. The suffix that describes an alkene is __________
7. The serial number of carbon atom at which a substituent is bonded is called __________
8. The suffix for a carboxylic group is __________
9. The molecule C_2H_5OH is called __________
10. The compound $CH_3 \cdot CO \cdot CH_3$ is a __________

F. Match the Columns Type Questions

IUPAC name		Trivial name		Formula	
1	Methane	A.	Acetylene	I.	HCHO
2	Ethane	B.	Formic acid	II.	C_2H_2
3	Ethene	C.	Formaldehyde	III.	CH_3COOH
4	Ethyne	D.	Acetic acid	IV.	HCOOH
5	Methanal	E.	Methane	V.	C_2H_6
6	Ethanal	F.	Ethane	VI.	C_2H_4
7	Methanoic acid	G.	Ethylene	VII.	CH_3CHO
8	Ethanoic acid	H.	Acetaldehyde	VIII.	CH_4

G. Multiple Choice Type Questions

Choose the correct alternative.

1. The compounds whose names end with "yne" have a __________ between any two carbon atoms.
   (a) ionic bond ❑ (b) single bond ❑ (c) double bond ❑ (d) triple bond ❑
2. The organic compounds containing the –OH group are called
   (a) acids ❑ (b) aldehydes ❑ (c) alcohols ❑ (d) esters ❑
3. Ethane (C_2H_6) has the following number of covalent bonds.
   (a) 6 ❑ (b) 7 ❑ (c) 8 ❑ (d) 9 ❑
4. Acetone is a three-carbon compound with the functional group
   (a) carboxylic ❑ (b) aldehydic ❑ (c) ketonic ❑ (d) alcoholic ❑

5. Alcohols are
   (a) neutral ❑ (b) basic ❑
   (c) acidic ❑ (d) amphoteric ❑
6. In alkenes, the two carbon atoms are joined by a/an
   (a) ionic bond ❑ (b) triple bond ❑
   (c) double bond ❑ (d) single bond ❑
7. 2-methylbutane is commonly known as
   (a) pentane ❑ (b) isopentane ❑
   (c) neopentane ❑ (d) none of these ❑
8. Saturated hydrocarbons are called
   (a) olefins ❑ (b) alkanes ❑
   (c) alkenes ❑ (d) alkynes ❑
9. The organic compounds containing –COOH group are called
   (a) alcohols ❑ (b) ketones ❑
   (c) esters ❑ (d) carboxylic acids ❑
10. The –CHO containing organic compound belongs to the category of
   (a) acids ❑ (b) aldehydes ❑
   (c) ketones ❑ (d) alcohols ❑

ANSWERS

A. 1. Ethene, Propene, Butene 2. Butane, C_4H_{10} 3. C_4H_6, C_2H_2
4. 2-chloro-2-methylpropane 5. Prop, ene 6. -one
7. Alkenes 8. Methanoic acid, Ethanoic acid
9. Functional group 10. Homologous series

D. 1. F 2. T 3. F 4. F 5. F 6. T 7. F 8. F 9. T 10. F

E. 1. Ethene 2. Hydrocarbons 3. Homologue 4. C_nH_{2n+2}
5. Alkanes 6. -ene 7. Locant 8. -oic acid
9. Ethanol 10. Propanone

F. 1 – E – VIII; 2 – F – V; 3 – G – VI; 4 – A – II 5 – C – I;
6 – H – VII; 7 – B – IV; 8 – D – III

G. 1. d 2. c 3. b 4. c 5. a 6. c 7. b 8. b 9. d 10. b.

Practice Assignment – 1

A. Balance the following chemical equations by Hit-and-Trial Method.

I. Reactions involving only one reactant

1. $KClO_3(s) \xrightarrow{heat} KCl(s) + O_2(g)$
2. $Pb_3O_4(s) \xrightarrow{heat} PbO(s) + O_2(g)$
3. $CaCO_3(s) \xrightarrow{heat} CaO(s) + CO_2(g)$
4. $HOCl(l) \longrightarrow HCl(l) + O_2(g)$
5. $H_2O_2(l) \longrightarrow H_2O(l) + O_2(g)$

II. Reactions involving two reactants

1. $P_4 + Cl_2 \rightarrow PCl_5$
2. $Fe + Cl_2 \rightarrow FeCl_3$
3. $FeCl_2 + Cl_2 \rightarrow FeCl_3$
4. $Cl_2 + H_2O \rightarrow HCl + O_2$
5. $Fe + H_2O(g) \xrightarrow{\Delta} Fe_3O_4 + H_2$
6. $Al + H_2O \rightarrow Al(OH)_3 + H_2$
7. $NH_3 + Cl_2 \rightarrow NH_4Cl + N_2$
8. $N_2H_4 + H_2O_2 \rightarrow N_2 + H_2O$
9. $Al_2(SO_4)_3 + NH_4OH \rightarrow Al(OH)_3 + (NH_4)_2SO_4$
10. $KI(aq) + Cl_2(aq) \rightarrow KCl(aq) + I_2(aq)$
11. $CH_4(g) + O_2(g) \rightarrow CO_2(g) + H_2O(l)$
12. $Na(s) + H_2O(l) \rightarrow NaOH(aq) + H_2(g)$
13. $NaOH + SO_3 \rightarrow Na_2SO_4 + H_2O$
14. $Al_2O_3 + NaOH \rightarrow NaAlO_2 + H_2O$
15. $MnO_2 + HCl \rightarrow MnCl_2 + H_2O + Cl_2$
16. $Na_2CO_3 + HCl \rightarrow NaCl + CO_2 + H_2O$
17. $PbO_2 + H_2SO_4 \rightarrow PbSO_4 + H_2O + O_2$
18. $NH_4Cl + Ca(OH)_2 \rightarrow CaCl_2 + NH_3 + H_2O$
19. $HNO_3 + H_2S \rightarrow NO_2 + H_2O + S$

20. $PbO_2 + HCl \rightarrow PbCl_2 + H_2O + Cl_2$
21. $Ag + HNO_3 \rightarrow AgNO_3 + NO_2 + H_2O$
22. $NaOH + Cl_2 \rightarrow NaCl + NaClO + H_2O$

III. Reactions involving more than two reactants

1. $SO_2 + H_2O + I_2 \rightarrow H_2SO_4 + HI$
2. $SO_2 + H_2O + Cl_2 \rightarrow H_2SO_4 + HCl$
3. $FeCl_3 + SO_2 + H_2O \rightarrow FeCl_2 + HCl + H_2SO_4$
4. $KI + H_2SO_4 + MnO_2 \rightarrow KHSO_4 + MnSO_4 + I_2 + H_2O$

B. Balance the following chemical equations by Partial Equation Method.

I. Reactions involving two reactants

1. $HNO_2 + H_2S \rightarrow S + H_2O + NO$
2. $HNO_2 + SO_2 \rightarrow H_2SO_4 + NO$
3. $HNO_2 + KI \rightarrow KOH + I_2 + NO$
4. $HNO_3 + S \rightarrow H_2SO_4 + NO_2 + H_2O$
5. $HNO_3 + C \rightarrow H_2CO_3 + NO_2 + H_2O$
6. $HNO_3 + P \rightarrow H_3PO_4 + NO_2 + H_2O$
7. $HNO_3 + I_2 \rightarrow HIO_3 + NO_2 + H_2O$
8. $HNO_3 + As \rightarrow H_3AsO_4 + NO_2 + H_2O$

II. Reactions involving three reactants

1. $HNO_2 + SnCl_2 + HCl \rightarrow SnCl_4 + H_2O + NO$
2. $FeSO_4 + H_2SO_4 + I_2 \rightarrow Fe_2(SO_4)_3 + HI$
3. $FeSO_4 + H_2SO_4 + Br_2 \rightarrow Fe_2(SO_4)_3 + HBr$

C. Rewrite the following word equations and balance the chemical equation by Hit-and-Trial Method.

1. Magnesium nitrate $\rightarrow$ Magnesium oxide + Nitrogen dioxide + Oxygen
2. Sodium thiosulphate + Iodine $\rightarrow$ Sodium iodide + Sodium tetrathionate
3. Nitric acid $\xrightarrow{\Delta}$ Nitrogen dioxide + Oxygen + Water
4. Aluminium hydroxide $\rightarrow$ Aluminium oxide + Water
5. Calcium + Water $\rightarrow$ Calcium hydroxide + Hydrogen
6. Lithium + Water $\rightarrow$ Lithium hydroxide + Hydrogen
7. Zinc + Sodium hydroxide $\rightarrow$ Sodium zincate + Hydrogen
8. Cuprous sulphide + Oxygen $\rightarrow$ Cuprous oxide + Sulphur dioxide
9. Hydrogen sulphide + Oxygen $\rightarrow$ Sulphur dioxide + Water
10. Calcium phosphide + Water $\rightarrow$ Calcium hydroxide + Phosphine

11. Ferric oxide + Carbon monoxide → Iron + Carbon dioxide
12. Magnesium + Sulphuric acid → Magnesium sulphate + Water + Sulphur dioxide
13. Burning of magnesium in air.
14. Magnesium reacts with sulphuric acid.

D. Rewrite the following word equations and balance the chemical equations by Partial Equation Method.

1. Lead sulphide + Ozone → Lead sulphate + Oxygen
2. Potassium sulphite + Hydrogen peroxide → Potassium sulphate + Water
3. Sulphuric acid + Hydriodic acid → Hydrogen sulphide + Iodine + Water
4. Potassium chlorate + Sodium sulphite → Potassium chloride + Sodium sulphate
5. Sodium carbonate + Sulphuric acid → Sodium sulphate + Water + Carbon dioxide
6. Aluminium + Sulphuric acid → Aluminium sulphate + Hydrogen
7. Ethylene + Hydrochloric acid → Chloroethane
8. Potassium iodide + Nitrogen peroxide → Potassium nitrite + Iodine

E. Complete the following word equations, rewrite the corresponding chemical equations and balance them.

1. Zinc carbonate + Sulphuric acid → ________ + ________ + Carbon dioxide
2. Magnesium oxide + Hydrochloric acid → ________ + ________
3. Ammonia + ________ → Ammonium nitrate
4. Zinc hydroxide + ________ → Zinc chloride + Water
5. ________ + Nitric acid → Copper(II) nitrate + Water + ________
6. Calcium oxide + ________ → Calcium chloride + ________
7. Sodium hydroxide + Sulphuric acid → ________ + Water
8. Sodium hydroxide + Hydrochloric acid → ________ + ________
9. Iron + ________ → Iron(II) chloride + Hydrogen

F. From the statements given below:

i. Identify the reactants and products.

ii. Write the chemical equations.

iii. Balance the chemical equations.

1. Zinc and hydrochloric acid react to give zinc chloride and hydrogen.
2. Sodium metal reacts with water to give sodium hydroxide and hydrogen gas.
3. Carbon dioxide passed through limewater gives white turbidity due to the formation of calcium carbonate.

4. Phosphorus burns in air to give phosphorus pentoxide.
5. Magnesium ribbon burns in air to give a white powder of magnesium oxide.
6. Lead nitrate on heating decomposes to give lead oxide, nitrogen dioxide and oxygen.
7. Potassium chlorate on heating decomposes to give potassium chloride and oxygen gas.
8. Methane burns in air to give carbon dioxide and water, and heat is evolved.
9. White precipitate of silver chloride is formed on adding a solution of sodium chloride to silver nitrate solution.

Some Common Arrows in Chemistry

Arrow	Name	Use
$\rightarrow$	Chemical reaction arrow	Indicates a reaction occurring from the reactants to the products
$\leftrightarrow$	Double-headed arrow	Used to separate two resonance structures
$\rightleftharpoons$	Equilibrium arrow	Shows the chemical reaction is a reversible reaction
$\uparrow\downarrow$	Up and down arrow	Represent electrons in orbital diagrams
$\uparrow$	Arrow facing up	Represents gaseous product in a chemical equation
$\downarrow$	Arrow facing down	Represents precipitate in a chemical equation
$\curvearrowleft$ $\curvearrowright$	Full-headed curved arrow	Shows the movement of an electron pair
⌒	Half-headed curved arrow (fishhook)	Shows the movement of a single electron
$\mapsto$	Dipole arrow	Shows direction of polarity in a bond (head of the arrow Points toward the more electronegative element)

Practice Assignment – 2

A. Balance the following chemical equations by Hit-and-Trial Method.

I. Reactions involving only one reactant

1. $H_2O(l) \xrightarrow{\text{electricity}} H_2(g) + O_2(g)$
2. $Pb(NO_3)_2(s) \xrightarrow{\text{heat}} PbO(s) + NO_2(g) + O_2(g)$
3. $ZnCO_3(s) \xrightarrow{\text{heat}} ZnO(s) + CO_2(g)$
4. $HgO(s) \xrightarrow{\text{heat}} Hg(l) + O_2(g)$
5. $NH_4Cl(s) \xrightarrow{\text{heat}} NH_3(g) + HCl(g)$

II. Reactions involving two reactants

1. $NaOH$ (hot) $+ Cl_2 \rightarrow NaCl + NaClO + H_2O$
2. $Al + Cl_2 \rightarrow AlCl_3$
3. $Zn + HCl \rightarrow ZnCl_2 + H_2$
4. $P_4 + O_2 \rightarrow P_2O_5$
5. $SO_2 + O_2 \rightarrow SO_3$
6. $PbO + O_2 \rightarrow Pb_3O_4$
7. $Fe + H_2O \rightarrow Fe_3O_4 + H_2$
8. $Na + H_2O \rightarrow NaOH + H_2$
9. $Al + H_2O \xrightarrow{\Delta} Al(OH)_3 + H_2$
10. $NO + H_2 \rightarrow N_2 + H_2O$
11. $Cu + NO \rightarrow CuO + N_2$
12. $FeS_2 + O_2 \rightarrow Fe_2O_3 + SO_2$
13. $H_2S_2O_7 + H_2O \rightarrow H_2SO_4$
14. $Mg + SO_2 \rightarrow MgO + S$
15. $H_2S + SO_2 \rightarrow H_2O + S$
16. $H_2S + O_2 \rightarrow H_2O + SO_2$
17. $FeS + HCl \rightarrow FeCl_2 + H_2S$

18. $PbS + O_2 \rightarrow PbO + SO_2$
19. $ZnS + O_2 \rightarrow ZnO + SO_2$
20. $ZnO + HCl \rightarrow ZnCl_2 + H_2O$
21. $MgO + HCl \rightarrow MgCl_2 + H_2O$
22. $Na_2O + H_2O \rightarrow NaOH$

III. Reactions involving three reactants

1. $SO_3 + H_2O + FeCl_3 \rightarrow FeCl_2 + HCl + H_2SO_4$
2. $Na_2SO_3 + H_2O + Cl_2 \rightarrow Na_2SO_4 + HCl$
3. $KMnO_4 + SO_2 + H_2O \rightarrow MnSO_4 + K_2SO_4 + H_2SO_4$
4. $Cl_2 + SO_2 + H_2O \rightarrow H_2SO_4 + HCl$

B. Balance the following chemical equations by Partial Equation Method.

I. Reactions involving two reactants

1. $HNO_3 + Sb \rightarrow H_3SbO_4 + NO_2 + H_2O$
2. $HNO_3 + Sn \rightarrow H_2SnO_3 + NO_2 + H_2O$
3. $HNO_3(\text{dil.}) + Zn \rightarrow Zn(NO_3)_2 + N_2O + H_2O$
4. $HNO_3(\text{v. dil.}) + Zn \rightarrow Zn(NO_3)_2 + NH_4NO_3 + H_2O$
5. $HNO_3(\text{conc.}) + Zn \rightarrow Zn(NO_3)_2 + NO_2 + H_2O$
6. $HNO_3(\text{dil.}) + Sn \rightarrow Sn(NO_3)_2 + NH_4NO_3 + H_2O$
7. $HNO_3(\text{dil.}) + Fe \rightarrow Fe(NO_3)_2 + NH_4NO_3 + H_2O$
8. $HNO_3(\text{conc.}) + Fe \rightarrow Fe(NO_3)_3 + NO_2 + H_2O$

II. Reactions involving three reactants

1. $FeSO_4 + H_2SO_4 + Cl_2 \rightarrow Fe_2(SO_4)_3 + HCl$
2. $FeSO_4 + H_2SO_4 + HNO_3 \rightarrow Fe_2(SO_4)_3 + H_2O + NO_2$
3. $FeSO_4 + H_2SO_4 + H_2O_2 \rightarrow Fe_2(SO_4)_3 + H_2O$

C. Rewrite the following word equations and balance the chemical equation by Hit-and-Trial Method.

1. Iron + Sulphuric acid → Ferric sulphate + Water + Sulphur dioxide
2. Ferric oxide + Carbon → Iron + Carbon monoxide
3. Aluminium + Hydrochloric acid → Aluminium chloride + Hydrogen
4. Silver nitrate + Sodium hydroxide → Sodium nitrate + Silver oxide + Water
5. Mercury + Sulphuric acid → Mercuric sulphate + Water + Sulphur dioxide
6. Aluminium hydroxide + Sulphuric acid → Aluminium sulphate + Water

7. Aluminium carbide + Water → Aluminium hydroxide + Methane
8. Magnesium nitride + Water → Magnesium hydroxide + Ammonia
9. Calcium hydroxide + Ammonium chloride → Calcium chloride + Ammonia + Water
10. Cuprous sulphide + Cuprous oxide → Copper + Sulphur dioxide
11. Manganese dioxide + Hydrochloric acid → Manganous chloride + Water + Chlorine
12. Potassium iodide + Hydrogen peroxide → Potassium hydroxide + Iodine
13. Sodium reacts with oxygen.
14. Sodium carbonate reacts with hydrochloric acid.

D. Rewrite the following word equations and balance the chemical equations by Partial Equation Method.

1. Nitrogen peroxide + Ozone → Nitrogen pentoxide + Oxygen
2. Nitrogen peroxide + Water → Nitrous acid + Nitric acid
3. Nitrogen peroxide + Sodium hydroxide → Sodium nitrite + Sodium nitrate + Water
4. Sulphur dioxide + Hydrogen sulphide → Water + Sulphur
5. Potassium manganate + Ozone + Water → Potassium permanganate + Potassium hydroxide + Oxygen
6. Hydrogen sulphide + Potassium permanganate + Sulphuric acid → Sulphur + Water + Potassium sulphate + Manganese sulphate
7. Potassium permanganate + Sulphuric acid + Hydrogen peroxide → Potassium sulphate + Manganese sulphate + Water + Oxygen
8. Potassium iodide + Ozone + Water → Potassium hydroxide + Oxygen + Iodine

E. Complete the following word equations, rewrite the corresponding chemical equations and balance them.

1. Sodium carbonate + ____________ → Sodium sulphate + Water
2. ____________ + Hydrochloric acid → Calcium chloride + Water
3. Magnesium oxide + ____________ → Magnesium nitrate + Water
4. ____________ + Sulphuric acid → Copper (II) sulphate + Water + Carbon dioxide
5. ____________ + Sulphuric acid → Magnesium sulphate + Hydrogen
6. Magnesium hydroxide + Hydrochloric acid → ____________ + Water
7. Copper(II) oxide + Hydrochloric acid → ____________ + Water
8. Aluminium hydroxide + Nitric acid → ____________ + Water
9. ____________ + Hydrochloric acid → Magnesium chloride + Water + Carbon dioxide

F. From the statements given below:

i. **Identify the reactants and products.**

ii. **Write the chemical equations.**

iii. **Balance the chemical equations.**

1. The colour of potassium permanganate gets decolourised when added to an acidified solution of ferrous sulphate.
2. Phosphorus reacts with chlorine to give phosphorus pentachloride.
3. Ferric oxide reacts with sulphuric acid to give ferric sulphate and water.
4. Sulphur trioxide reacts with water to give sulphuric acid.
5. Magnesium nitride reacts with water to produce magnesium hydroxide and ammonia.
6. Magnesium burns in carbon dioxide to form magnesium oxide and carbon.
7. Carbon disulphide burns in air to give carbon dioxide and sulphur dioxide.
8. Chlorine gas burns in hydrogen gas to give hydrogen chloride.
9. Barium chloride reacts with zinc sulphate to give zinc chloride and a precipitate of barium sulphate.

Practice Assignment – 3

A. Balance the following chemical equations by Hit-and-Trial Method.

I. Reactions involving only one reactant

1. $FeSO_4(s) \xrightarrow[\text{strongly}]{\text{heat}} Fe_2O_3(s) + SO_2(g) + SO_3(g)$
2. $NH_3(g) \rightarrow N_2(g) + H_2(g)$
3. $Cu(NO_3)_2(s) \xrightarrow{\Delta} CuO(s) + NO_2(g) + O_2(g)$
4. $H_2CO_3 \xrightarrow{\Delta} H_2O + CO_2$

II. Reactions involving two reactants

1. $NaOH + H_2S \rightarrow Na_2S + H_2O$
2. $NH_4OH + H_2S \rightarrow (NH_4)_2S + H_2O$
3. $SbCl_3 + H_2S \rightarrow Sb_2S_3 + HCl$
4. $NH_3 + Cl_2 \rightarrow NH_4Cl + N_2$
5. $FeCl_3 + NH_4OH \rightarrow Fe(OH)_3 + NH_4Cl$
6. $Al_2(SO_4)_3 + NH_4OH \rightarrow Al(OH)_3 + (NH_4)_2SO_4$
7. $Ca_3(PO_4)_2 + H_3PO_4 \rightarrow Ca(H_2PO_4)_2$
8. $ZnO + NaOH \rightarrow Na_2ZnO_2 + H_2O$
9. $Al_2O_3 + H_2SO_4 \rightarrow Al_2(SO_4)_3 + H_2O$
10. $CO_2 + NaOH \rightarrow Na_2CO_3 + H_2O$
11. $Fe_3O_4 + HCl \rightarrow FeCl_3 + FeCl_2 + H_2O$
12. $C + HNO_3 \rightarrow CO_2 + NO_2 + H_2O$
13. $HNO_3 + H_2S \rightarrow NO_2 + H_2O + S$
14. $Cu + H_2SO_4 \rightarrow CuSO_4 + H_2O + SO_2$
15. $NH_4Cl + Ca(OH)_2 \rightarrow CaCl_2 + NH_3 + H_2O$
16. $NH_3 + CuO \rightarrow Cu + H_2O + N_2$
17. $NH_4OH + H_2SO_4 \rightarrow (NH_4)_2SO_4 + H_2O$
18. $Na_2CO_3 + HNO_3 \rightarrow NaNO_3 + H_2O + CO_2$
19. $P_4 + HNO_3 \rightarrow H_3PO_4 + NO_2 + H_2O$

20. $P_4 + H_2SO_4 \rightarrow H_3PO_4 + SO_2 + H_2O$
21. $NaCl + H_2SO_4 \rightarrow Na_2SO_4 + HCl$

III. Reactions involving three reactants

1. $KNO_2 + H_2O + Cl_2 \rightarrow KNO_3 + HCl$
2. $Na_3AsO_3 + H_2O + Cl_2 \rightarrow Na_3AsO_4 + HCl$
3. $Li + CO_2 + H_2O \rightarrow LiHCO_3 + H_2$
4. $Al + NaOH + H_2O \rightarrow NaAl(OH)_4 + H_2$

B. Balance the following chemical equations by Partial Equation Method.

I. Reactions involving two reactants

1. HNO_3(dil.) + Cu $\rightarrow Cu(NO_3)_2 + H_2O + NO$
2. HNO_3(conc.) + Cu $\rightarrow Cu(NO_3)_2 + H_2O + NO_2$
3. HNO_3(dil.) + Pb $\rightarrow Pb(NO_3)_2 + NO + H_2O$
4. HNO_3(conc.) + Pb $\rightarrow Pb(NO_3)_2 + NO_2 + H_2O$
5. HNO_3(dil.) + Ag $\rightarrow AgNO_3 + NO + H_2O$
6. HNO_3(conc.) + Hg $\rightarrow Hg(NO_3)_2 + NO_2 + H_2O$
7. HNO_3(dil.) + Hg $\rightarrow Hg(NO_3)_2 + NO + H_2O$

II. Reactions involving three reactants

1. $FeSO_4 + H_2SO_4 + O_3 \rightarrow Fe_2(SO_4)_3 + H_2O$
2. $FeSO_4 + H_2SO_4 + K_2Cr_2O_7 \rightarrow Fe_2(SO_4)_3 + Cr_2(SO_4)_3 + K_2SO_4 + H_2O$
3. $FeSO_4 + H_2SO_4 + K_2CrO_4 \rightarrow Fe_2(SO_4)_3 + Cr_2(SO_4)_3 + K_2SO_4 + H_2O$
4. $FeSO_4 + H_2SO_4 + KMnO_4 \rightarrow Fe_2(SO_4)_3 + MnSO_4 + K_2SO_4 + H_2O$

C. Rewrite the following word equations and balance the chemical equation by Hit-and-Trial Method.

1. Ferrous sulphate + Sulphuric acid + Oxygen → Ferric sulphate + Water
2. Lead + Water + Oxygen → Lead(II) hydroxide
3. Sodium meta-aluminate + Water + Carbon dioxide → Aluminium hydroxide + Sodium carbonate
4. Tin + Water + Sodium hydroxide → Sodium stannate + Hydrogen
5. Aluminium + Water + Sodium hydroxide → Sodium meta-aluminate + Hydrogen
6. Copper(II) carbonate + Hydrochloric acid → Copper(II) chloride + Water + Carbon dioxide
7. Lead(II) nitrate + Potassium iodide → Lead(II) iodide + Potassium nitrate
8. Potassium hydroxide + Sulphuric acid → Potassium sulphate + Water
9. Magnesium + Copper(II) sulphate → Copper + Magnesium sulphate

10. Reaction between hydrochloric acid and sodium hydroxide.
11. Calcium carbonate decomposes to produce calcium oxide and carbon dioxide gas.

D. Rewrite the following word equations and balance the chemical equations by Partial Equation Method.

1. Potassium permanganate + Sulphuric acid + Oxalic acid → Potassium sulphate + Manganese sulphate + Carbon dioxide + Water
2. Potassium iodide + Cupric sulphate + Sodium thiosulphate → Sodium persulphate + Cuprous iodide + Potassium sulphate + Sodium iodide
3. Hydrogen sulphide + Chlorine + Water → Hydrochloric acid + Sulphur + Water
4. Sodium sulphite + Chlorine + Water → Sodium sulphate + Hydrochloric acid + Water
5. Potassium chlorate + Potassium iodide + Hydrochloric acid → Potassium chloride + Iodine + Water
6. Potassium permanganate + Sulphuric acid + Nitric oxide → Potassium sulphate + Manganese sulphate + Nitric acid
7. Sulphur dioxide + Nitrogen peroxide + Water → Sulphuric acid + Nitric oxide
8. Potassium dichromate + Sulphur dioxide + Sulphuric acid → Potassium sulphate + Chromium sulphate + Water
9. Aluminium + Water + Sodium hydroxide → Trisodium aluminium hydroxide + Hydrogen

E. Complete the following word equations, rewrite the corresponding chemical equations and balance them.

1. Sodium hydrogencarbonate + Hydrochloric acid → ________ + ________ + ________
2. Aluminium + Hydrochloric acid → ________ + Hydrogen
3. Magnesium + Nitric acid → ________ + Hydrogen
4. Zinc oxide + ________ → Zinc chloride + Water
5. ________ + Nitric acid → Calcium nitrate + Water + Carbon dioxide
6. Zinc hydroxide + Sulphuric acid → ________ + ________
7. Aluminium + ________ → Aluminium sulphate +Hydrogen
8. Magnesium hydroxide + ________ → Magnesium sulphate + ________
9. Copper(II) carbonate + Sulphuric acid → ________ + Water + Carbon dioxide
10. Zinc + Sulphuric acid → Zinc sulphate + ________

F. From the statements given below:

i. Identify the reactants and products.

ii. Write the chemical equations.

iii. Balance the chemical equations.

1. Hydrogen sulphide gas burns in air to give water and sulphur dioxide.
2. Aluminium metal displaces iron from iron(III) oxide giving aluminium oxide and iron.
3. Hydrogen gas combines with nitrogen to give ammonia.
4. Barium chloride reacts with aluminium sulphate to give aluminium chloride and precipitate of barium sulphate.
5. Potassium metal reacts with water to give potassium hydroxide and hydrogen gas.
6. Aqueous solutions of sulphuric acid and sodium hydroxide react to form aqueous sodium sulphate and water.
7. Hydrogen sulphide is passed into an aqueous solution of sulphur dioxide to form sulphur and water.
8. Zinc carbonate is heated to form zinc oxide and carbon dioxide.
9. Ammonia gas is passed over heated copper(II) oxide.

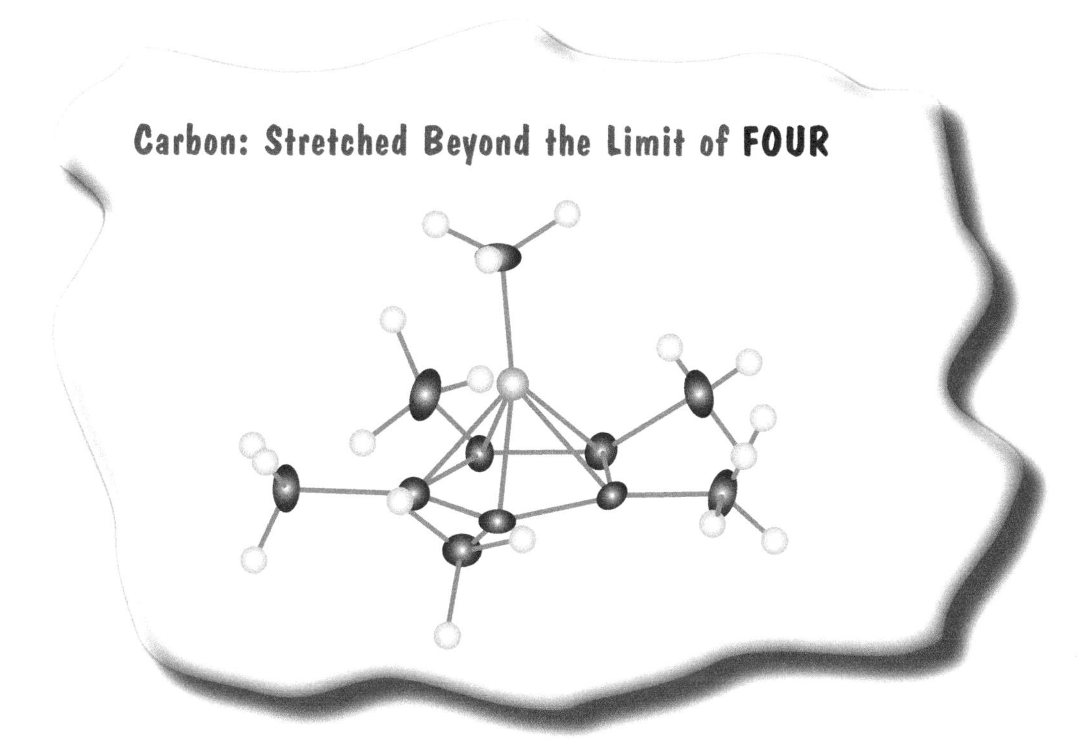

https://www.sciencenews.org/article/carbon-can-exceed-four-bond-limit

Practice Assignment – 4

A. Oxidation Numbers and their Use

1. Write down the oxidation number of each element in the following compounds.

(a) Al_2O_3 (b) $AlCl_3$ (c) $NaCl$ (d) Al_2Se_3

(e) CH_4 (f) CH_3Cl (g) CH_2Cl_2 (h) $CHCl_3$

(i) CCl_4 (j) ZnO (k) K_2PtCl_6 (l) $H_2C_2O_4$

(m) KI (n) MnO_2 (o) $KMnO_4$ (p) K_2CrO_4

(q) $K_2Cr_2O_7$ (r) $NaBrO$ (s) K_2SO_4 (t) $HClO$

(u) I_2 (v) N_2O_5 (w) HNO_3 (x) N_2O_3

(y) $HClO_4$ (z) H_3PO_4

2. Write down the oxidation number of each element in the following ions.

(a) CH_3COO^- (b) PO_4^{3-} (c) CrO_4^{2-} (d) HPO_3^{2-}

(e) NO_3^- (f) SO_4^{2-} (g) SO_3^{2-} (h) S^{2-}

(i) $C_2O_4^{2-}$ (j) ClO^- (k) ClO_4^- (l) Fe^{3+}

(m) Al^{3+} (n) Cu^{2+} (o) Hg^{2+} (p) La^{3+}

(q) NO_2^- (r) CO_3^{2-} (s) BO_3^{3-} (t) Mg^{2+}

(u) Ca^{2+}

3. Indicate the oxidation number of the atoms with Bold face.

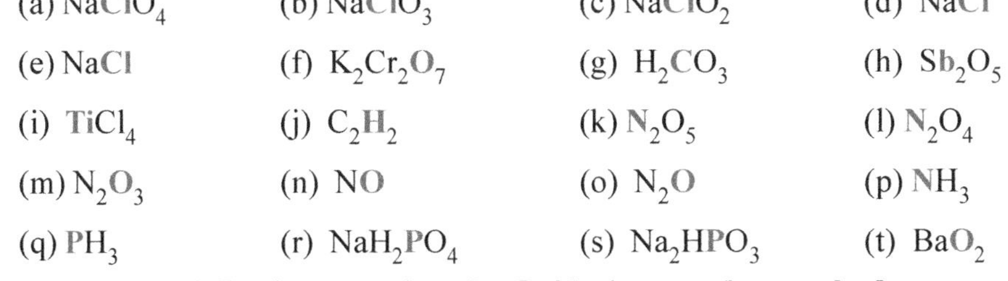

(a) $NaClO_4$ (b) $NaClO_3$ (c) $NaClO_2$ (d) $NaCl$

(e) $NaCl$ (f) $K_2Cr_2O_7$ (g) H_2CO_3 (h) Sb_2O_5

(i) $TiCl_4$ (j) C_2H_2 (k) N_2O_5 (l) N_2O_4

(m) N_2O_3 (n) NO (o) N_2O (p) NH_3

(q) PH_3 (r) NaH_2PO_4 (s) Na_2HPO_3 (t) BaO_2

4. Balance the following equations by Oxidation number method.

(a) $Sn + Cl_2 \rightarrow SnCl_2$

(b) $H_2S + HNO_3 \rightarrow H_2O + NO + S$

(c) $CuS + HNO_3 \rightarrow Cu(NO_3)_2 + NO + S + H_2O$

(d) $Al + HNO_3 \rightarrow Al(NO_3)_3 + N_2O + H_2O$

(e) $KMnO_4 + KBr + H_2SO_4 \rightarrow Br_2 + K_2SO_4 + MnSO_4 + H_2O$

(f) $Al^{3+} + OH^- \rightarrow Al(OH)_4^-$

(g) $H_2O_2 + Br_2 \rightarrow BrO_3^- + H_2O$

(h) $P + HNO_3 \rightarrow H_3PO_4 + NO$

(i) $KMnO_4 + C_2O_4H_2 + H_2SO_4 \rightarrow K_2SO_4 + MnSO_4 + CO_2 + H_2O$

(j) $KMnO_4 + H_2SO_4 + C_2H_5OH \rightarrow K_2SO_4 + MnSO_4 + CH_3COOH + H_2O$

(k) $HClO \rightarrow HCl + HClO_3$

(l) $Cu + HNO_3 \rightarrow Cu(NO_3)_2 + NO + H_2O$

(m) $Mg + HNO_3 \rightarrow Mg(NO_3)_2 + NH_4NO_3 + H_2O$

(n) $Cu + H_2SO_4 \rightarrow CuSO_4 + SO_2 + H_2O$

(o) $KMnO_4 + HCl \rightarrow KCl + MnCl_2 + H_2O + Cl_2$

5. Balance the following equations and indicate in each case the oxidising agent, the reducing agent, the substance oxidised and the substance reduced.

(a) $H_2S + H_2SO_3 \rightarrow H_2O + S$

(b) $Fe + H_2SO_4 \rightarrow FeSO_4 + H_2$

(c) $C + H_2SO_4 \rightarrow CO_2 + SO_2 + H_2O$

(d) $Al + H_2SO_4 \rightarrow Al_2(SO_4)_3 + SO_2 + H_2O$

B. Writing Ionic Equations and their Balancing

I. Rewrite the following equations in the ionic form and balance them using Hit-and-Trial method.

1. $NH_4Cl + NaOH \rightarrow NH_3 + NaCl + H_2O$
2. $CuSO_4 + H_2S \rightarrow CuS + H_2SO_4$
3. $Bi(OH)_4 + HCl \rightarrow BiCl_3 + H_2O$
4. $BaCl_2 + Al_2(SO_4)_3 \rightarrow BaSO_4 + AlCl_3$
5. $FeSO_4 + KCNS \rightarrow Fe(CNS)_2 + K_2SO_4$
6. $AgNO_3 + HBr \rightarrow AgBr(s) + HNO_3$
7. $AgNO_3 + KI \rightarrow AgI(s) + KNO_3$
8. $(CH_3COO)_2Pb + H_2S \rightarrow PbS(s) + CH_3COOH$
9. $NaNO_3 + H_2SO_4 \rightarrow Na_2SO_4 + H_2O + NO_2 + O_2$

10. $Cu + H_2SO_4 + HNO_3 \rightarrow CuSO_4 + H_2O + NO_2$
11. $NH_3 + HCl \rightarrow NH_4Cl$
12. $HCl + NaOH \rightarrow NaCl + H_2O$
13. $BiCl_3 + H_2S \rightarrow Bi_2S_3(s) + HCl$
14. $AlCl_3 + NH_4OH \rightarrow Al(OH)_3(s) + NH_4Cl$
15. $Ca(NO_3)_2 + (NH_4)_2SO_4 \rightarrow CaSO_4(s) + NH_4NO_3$
16. $CH_3COOH + NaOH \rightarrow CH_3COONa + H_2O$
17. $HCl + Na_2CO_3 \rightarrow NaHCO_3 + NaCl$
18. $HCl + NaHCO_3 \rightarrow NaCl + H_2O + CO_2$
19. $Na_2B_4O_7 + H_2O \rightarrow NaOH + H_3BO_3$
20. $FeCl_3 + SnCl_2 \rightarrow FeCl_2 + SnCl_4$
21. $KMnO_4 + FeSO_4 + H_2SO_4 \rightarrow K_2SO_4 + MnSO_4 + Fe_2(SO_4)_3 + H_2O$
22. $KMnO_4 + (COOH)_2 + H_2SO_4 \rightarrow K_2SO_4 + MnSO_4 + CO_2 + H_2O$
23. $Na_2S_2O_3 + I_2 \rightarrow Na_2S_4O_6 + NaI$
24. $MgSO_4 + KF \rightarrow MgF_2 + K_2SO_4$
25. $CaBr_2 + NaF \rightarrow CaF_2 + NaBr$
26. $NaClO_4 + MgCl_2 \rightarrow Mg(ClO_4)_2 + NaCl$
27. $K_2Cr_2O_7 + FeSO_4 + H_2SO_4 \rightarrow K_2SO_4 + Cr_2(SO_4)_3 + Fe_2(SO_4)_3 + H_2O$
28. $K_2CrO_4 + (COOH)_2 + H_2SO_4 \rightarrow K_2SO_4 + Cr_2(SO_4)_3 + CO_2 + H_2O$
29. $K_2Cr_2O_7 + SO_2 + H_2SO_4 \rightarrow K_2CrO_4 + Cr_2(SO_4)_3 + H_2SO_4$
30. $NaClO_4 + KI + HCl \rightarrow NaCl + KCl + H_2O + I_2$
31. $NaClO_3 + KNO_2 + H_2SO_4 \rightarrow KNO_3 + NaCl + H_2SO_4$
32. $KMnO_4 + KNO_2 + H_2SO_4 \rightarrow K_2SO_4 + KNO_3 + MnSO_4 + H_2O$
33. $Ag + HNO_2 + HNO_3 \rightarrow AgNO_3 + NO + H_2O$
34. $As + HNO_3 + H_2O \rightarrow H_3AsO_3 + NO$

II. Rewrite the following equations in the ionic form and balance them using Hit-and-Trial method.

1. $Na_2CO_3 + H_2SO_4 \rightarrow Na_2SO_4 + H_2O + CO_2$
2. $Ca(OH)_2 + CO_2 \rightarrow CaCO_3 + H_2O$
3. $CH_3COONa + H_2SO_4 \rightarrow CH_3COOH + Na_2SO_4$
4. $FeS + H_2SO_4 \rightarrow FeSO_4 + H_2S$
5. $(NH_4)_2SO_3 + H_2SO_4 \rightarrow (NH_4)_2SO_4 + SO_2 + H_2O$
6. $Cu + KNO_3 + H_2SO_4 \rightarrow K_2SO_4 + H_2O + NO_2 + CuSO_4$
7. $KCl + H_2SO_4 \rightarrow KHSO_4 + Cl_2 + H_2$
8. $KBr + H_2SO_4 \rightarrow KHSO_4 + Br_2 + H_2$
9. $KI + H_2SO_4 \rightarrow KHSO_4 + I_2 + H_2$
10. $Na_2SO_4 + BaCl_2 \rightarrow BaSO_4 + NaCl$
11. $KF + SiO_2 + H_2SO_4 \rightarrow SiF_4 + K_2SO_4 + H_2O$
12. $CuSO_4 + CaCl_2 \rightarrow CaSO_4 + CuCl_2$
13. $CuSO_4 + PbCl_2 \rightarrow PbSO_4 + CuCl_2$
14. $MgSO_4 + BaCl_2 \rightarrow BaSO_4 + MgCl_2$
15. $HNO_3 + BaCl_2 \rightarrow Ba(NO_3)_2 + HCl$
16. $HCl + HNO_3 \rightarrow NOCl + H_2O + [O]$
17. $PCl_5 + NaOCl \rightarrow POCl_3 + NaCl + Cl_2$
18. $KI + Ag + H_2O \rightarrow AgI + KOH + H_2$

III. Rewrite the following equations in the ionic form and balance them by Partial equation method.

1. $Zn + H_2SO_4 \rightarrow Zn^{2+} + SO_4^{2-} + H_2$
2. $Cl_2 + Br^- \rightarrow Br_2 + Cl^-$
3. $H^+ + Fe^{2+} + MnO_4^- \rightarrow Fe^{3+} + Mn^{2+} + H_2O$
4. $Cr_2O_7^{2-} + SO_2 + H^+ \rightarrow Cr^{3+} + HSO_4^- + H_2O$
5. $Cu + NO_3^- \rightarrow Cu^{2+} + NO$ (acid solution)
6. $NF_3 + H_2O \rightarrow HF + NO + NO_2$
7. $BrO_3^- + Br^- \rightarrow Br_2 + H_2O$
8. $I_2 + Cl_2 + H_2O \rightarrow HIO_3 + HCl$
9. $Ag + I^- + H_2O \rightarrow AgI + OH^- + H_2$ (acid solution)
10. $MnO_4^- + Cl^- \rightarrow MnO_2 + Cl_2$ (acid solution)
11. $Ag_2O + HCHO \rightarrow Ag + HCOO^-$ (basic solution)

12. $H_2SO_3 + Fe^{3+} \rightarrow Fe^{2+} + SO_4^{2-}$
13. $HNO_2 + Fe^{3+} \rightarrow Fe^{2+} + NO_3^-$
14. $Sn^{2+} + Fe^{3+} \rightarrow Sn^{4+} + Fe^{2+}$
15. $BaO_2 + Cl^- \rightarrow Ba^{2+} + Cl_2$ (acid solution)
16. $CrO_2^- + ClO^- \rightarrow Cl^- + CrO_4^{2-}$
17. $CO_3^{2-} + H_2SO_4 \rightarrow CO_2 + H_2O + SO_4^{2-}$
18. $C_2O_4^{2-} + MnO_4^- \rightarrow Mn^{2+} + CO_2$
19. $MnO_4^- + I^- + H^+ \rightarrow Mn^{2+} + I_2 + H_2O$
20. $Cr_2O_7^{2-} + I^- + H^+ \rightarrow Cr^{3+} + I_2 + H_2O$
21. $IO_3^- + I^- + H^+ \rightarrow I_2 + H_2O$
22. $BrO_3^- + I^- + H^+ \rightarrow Br^- + I_2 + H_2O$
23. $Ce^{4+} + I^- \rightarrow Ce^{3+} + I_2$
24. $Fe^{3+} + I^- \rightarrow Fe^{2+} + I_2$
25. $H_2O_2 + I^- + H^+ \xrightarrow{Mo} H_2O + I_2$
26. $H_3AsO_4 + I^- + H^+ \rightarrow H_3AsO_3 + I_2 + H_2O$
27. $Cu^{2+} + I^- \rightarrow Cu^+ + I_2$
28. $HNO_2 + I^- \rightarrow I_2 + NO + H_2O$
29. $SeO_3^{2-} + I^- + H^+ \rightarrow Se + I_2 + H_2O$
30. $O_3 + I^- + H^+ \rightarrow O_2 + I_2 + H_2O$
31. $Cl_2 + I^- \rightarrow Cl^- + I_2$
32. $Br_2 + I^- \rightarrow Br^- + I_2$
33. $HClO + I^- + H^+ \rightarrow Cl^- + I_2 + H_2O$
34. $Cr_2O_7^{2-} + Fe^{2+} \rightarrow Cr^{3+} + Fe^{3+}$
35. $CrO_4^{2-} + Fe^{2+} \rightarrow Fe^{3+} + Cr^{3+}$
36. $Hg^{2+} + Sn^{2+} \rightarrow Hg_2^{2+} + Sn^{4+}$
37. $ClO^- + SO_3^{2-} \rightarrow Cl^- + SO_4^{2-}$
38. $ClO_3^- + Fe^{2+} \rightarrow Cl^- + Fe^{3+}$
39. $Cu^{2+} + I^- \rightarrow Cu^+ + I_2$
40. $S_2O_3^{2-} + I_2 \rightarrow S_4O_6^{2-} + I^-$
41. $Sn^{2+} + Cr_2O_7^{2-} \rightarrow Sn^{4+} + Cr^{3+}$
42. $I_2 + NO_3^- \rightarrow IO_3^- + NO_2$
43. $I^- + Cl_2 \rightarrow I_2 + Cl^-$
44. $OH^- + Ca^{2+} + CO_2 \rightarrow CaCO_3 + H_2O$
45. $HCO_3^- \rightarrow CO_2 + H_2O$ (acid solution)

NAMES AND FORMULAE OF SOME IMPORTANT ORGANIC COMPOUNDS

Saturated hydrocarbons: Alkanes

General formula C_nH_{2n+2}

Formula		Name	
Molecular	**Condensed**	**Common**	**IUPAC**
CH_4	CH_4	Methane	Methane
C_2H_6	CH_3-CH_3	Ethane	Ethane
C_3H_8	$CH_3-CH_2-CH_3$	Propane	Propane
C_4H_{10}	$CH_3-CH_2-CH_2-CH_3$	Butane	Butane
C_5H_{12}	$CH_3-CH_2-CH_2-CH_2-CH_3$	Pentane	Pentane

Unsaturated hydrocarbons: Alkenes

General formula: C_nH_{2n}

Formula		Name	
Molecular	**Condensed**	**Common**	**IUPAC**
C_2H_4	$CH_2=CH_2$	Ethylene	Ethene
C_3H_6	$CH_3-CH=CH_2$	Propylene	Propene
C_4H_8	$CH_2=CH-CH_2-CH_3$	Butylene	Butene
C_4H_8	$CH_3-CH=CH-CH_3$	Butyl-2-ene	But-2-ene

Unsaturated hydrocarbons: Alkynes

General formula: C_nH_{2n-2}

Formula		Name	
Molecular	**Condensed**	**Common**	**IUPAC**
C_2H_2	$CH\equiv CH$	Acetylene	Ethyne
C_3H_4	$CH_3-C\equiv CH$	Methylacetylene (allylene)	Propyne
C_4H_6	$CH_3-C\equiv C-CH_3$	Dimethylacetylene	But-2-yne
C_4H_6	$CH_3-CH_2-C\equiv CH$	Ethylacetylene	Butyne

Alcohol: Alkanol

General formula: $C_nH_{2n+1}OH$

Formula		Name	
Molecular	**Condensed**	**Common**	**IUPAC**
CH_3OH	CH_3OH	Methyl alcohol	Methanol
C_2H_5OH	CH_3-CH_2OH	Ethyl alcohol	Ethanol
C_3H_7OH	$CH_3-CH_2-CH_2OH$	*n*-Propyl alcohol	Propan-l-ol (or Propanol)
C_3H_7OH	$CH_3-\underset{\mid \atop OH}{CH}-CH_3$	*iso*-Propyl alcohol	Propan-2-ol

Carboxylic acid: Alkanoic acid

General formula: $C_nH_{2n+1}COOH$

Formula		Name	
Molecular	**Condensed**	**Common**	**IUPAC**
$HCOOH$	$HCOOH$	Formic acid	Methanoic acid
CH_3COOH	CH_3COOH	Acetic acid	Ethanoic acid
C_2H_5COOH	CH_3-CH_2COOH	Propionic acid	Propanoic acid
C_3H_7COOH	$CH_3-CH_2-CH_2COOH$	Butyric acid	Butanoic acid
$C_{17}H_{35}COOH$	$CH_3-(CH_2)_{16}-COOH$	Stearic acid	Octadecanoic acid

ATOMIC MASSES OF ELEMENTS (BASED ON $^{12}_{6}C$ SCALE)

NAME	SYMBOL	ATOMIC NUMBER	ATOMIC MASS
Actinium	Ac	89	(227)
Aluminium	Al	13	26.9815
Americium	Am	95	(243)
Antimony	Sb	51	121.75
Argon	Ar	18	39.948
Arsenic	As	33	74.9216
Astatine	At	85	(210)
Barium	Ba	56	137.34
Berkelium	Bk	97	(247)
Beryllium	Be	4	9.0122
Bismuth	Bi	83	208.980
Bohrium	Bh	107	(262)
Boron	B	5	10.811
Bromine	Br	35	79.909
Cadmium	Cd	48	112.40
Calcium	Ca	20	40.08
Californium	Cf	98	(251)
Carbon	C	6	12.01115
Cerium	Ce	58	140.12
Cesium	Cs	55	132.905
Chlorine	Cl	17	35.453
Chromium	Cr	24	51.996
Cobalt	Co	27	58.9332
Copper	Cu	29	63.54
Curium	Cm	96	(247)
Dubnium	Db	105	(262)
Dysprosium	Dy	66	162.50
Einsteinium	Es	99	(254)
Erbium	Er	68	167.93
Europium	Eu	63	151.96
Fermium	Fm	100	(253)
Fluorine	F	9	18.9984
Francium	Fr	87	(233)
Gadolinium	Gd	64	157.25
Gallium	Ga	31	69.72
Germanium	Ge	32	72.59
Gold	Au	79	196.967
Hafnium	Hf	72	178.49
Hassium	Hs	108	(265)
Helium	He	2	4.0026
Holmium	Ho	67	164.930
Hydrogen	H	1	1.00797
Indium	In	49	114.82
Iodine	I	53	126.904
Iridium	Ir	77	192.2
Iron	Fe	26	55.847
Krypton	Kr	36	83.80
Lanthanum	La	57	138.91
Lawrencium	Lr	103	(257)
Lead	Pb	82	207.19
Lithium	Li	3	6.939
Lutetium	Lu	71	174.97
Magnesium	Mg	12	24.312
Manganese	Mn	25	54.9380
Meitnerium	Mt	109	(266)
Mendelevium	Md	101	(256)

NAME	SYMBOL	ATOMIC NUMBER	ATOMIC MASS
Mercury	Hg	80	200.59
Molybdenum	Mo	42	95.94
Moscovium	Mc	115	(288)
Neodymium	Nd	60	141.24
Neon	Ne	10	20.183
Neptunium	Np	93	(237)
Nickel	Ni	28	58.71
Nihonium	Nh	113	(284)
Niobium	Nb	41	92.906
Nitrogen	N	7	14.0467
Nobelium	No	102	(254)
Oganesson	Og	118	(294)
Osmium	Os	76	190.2
Oxygen	O	8	15.9994
Palladium	Pd	46	106.4
Phosphorus	P	15	30.9734
Platinum	Pt	78	195.09
Plutonium	Pu	94	(242)
Polonium	Po	84	(210)
Potassium	K	19	39.102
Praseodymium	Pr	59	104.908
Promethium	Pm	61	(145)
Protactinium	Pa	91	(231)
Radium	Ra	88	(226)
Radon	Rn	86	(222)
Rhenium	Re	75	186.2
Rhodium	Rh	45	102.905
Rubidium	Rb	37	85.47
Rutherfordium	Rf	104	(261)
Ruthenium	Ru	44	101.07
Samarium	Sm	62	150.35
Scandium	Sc	21	44.946
Seaborgium	Sg	106	(263)
Selenium	Se	34	78.96
Silicon	Si	14	28.086
Silver	Ag	47	107.870
Sodium	Na	11	22.9898
Strontium	Sr	38	87.62
Sulphur	S	16	32.064
Tantalum	Ta	73	180.94
Technetium	Tc	43	(99)
Tellurium	Te	51	127.604
Tennessine	Ts	117	(294)
Terbium	Tb	65	158.93
Thallium	Tl	81	204.37
Thorium	Th	90	232.038
Thulium	Tm	69	168.93
Tin	Sn	50	118.69
Titanium	Ti	22	47.90
Tungsten	W	74	183.85
Uranium	U	92	238.03
Vanadium	V	23	50.942
Xenon	Xe	54	131.30
Ytterbium	Yb	70	173.04
Yttrium	Y	39	88.905
Zinc	Zn	30	65.38
Zirconium	Zr	40	91.22

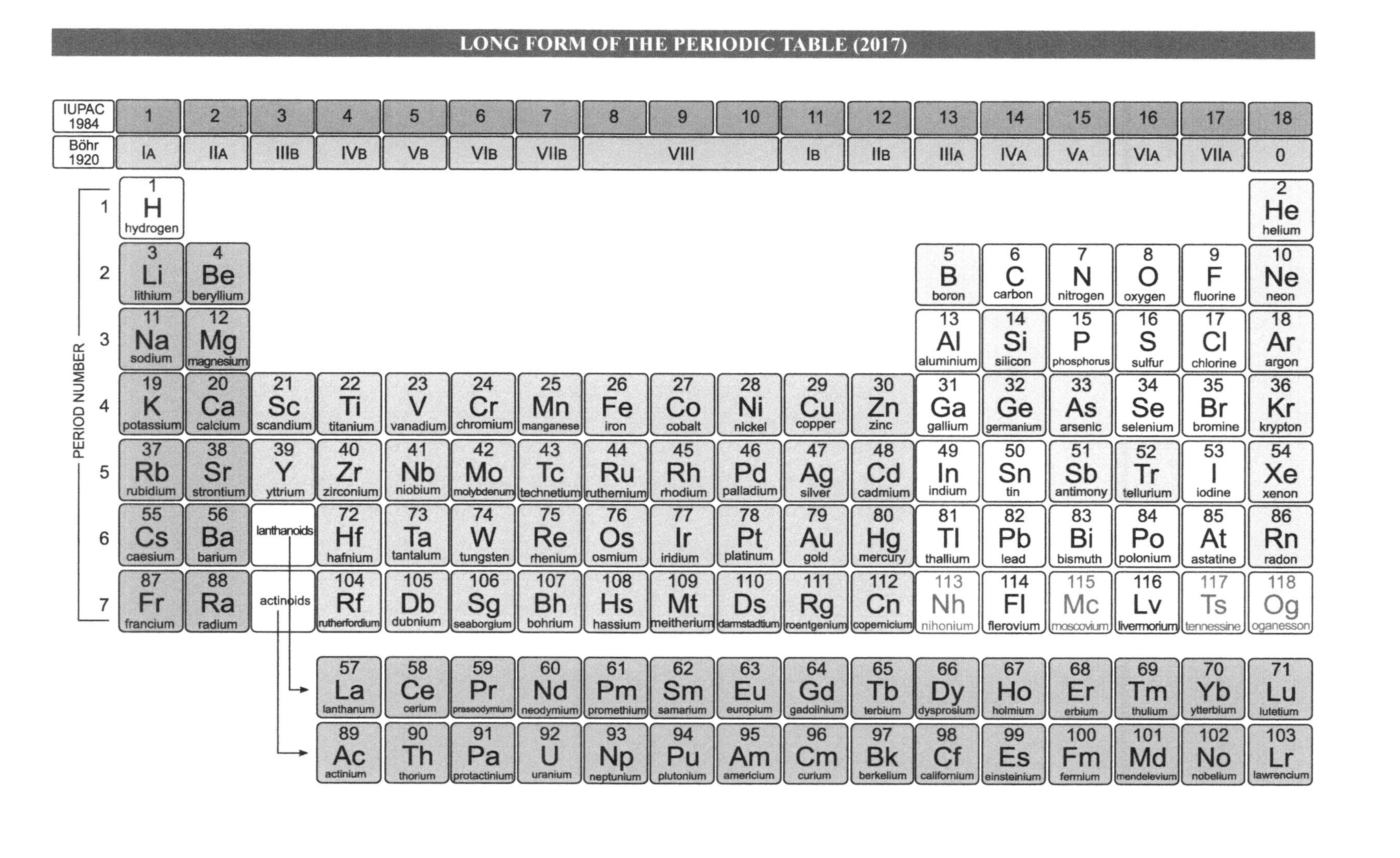

LONG FORM OF THE PERIODIC TABLE (2017)

IUPAC 1984	1	2	3	4	5	6	7	8	9	10	11	12	13	14	15	16	17	18
Böhr 1920	IA	IIA	IIIB	IVB	VB	VIB	VIIB	VIII			IB	IIB	IIIA	IVA	VA	VIA	VIIA	0
PERIOD NUMBER 1	1 H hydrogen																	2 He helium
2	3 Li lithium	4 Be beryllium											5 B boron	6 C carbon	7 N nitrogen	8 O oxygen	9 F fluorine	10 Ne neon
3	11 Na sodium	12 Mg magnesium											13 Al aluminium	14 Si silicon	15 P phosphorus	16 S sulfur	17 Cl chlorine	18 Ar argon
4	19 K potassium	20 Ca calcium	21 Sc scandium	22 Ti titanium	23 V vanadium	24 Cr chromium	25 Mn manganese	26 Fe iron	27 Co cobalt	28 Ni nickel	29 Cu copper	30 Zn zinc	31 Ga gallium	32 Ge germanium	33 As arsenic	34 Se selenium	35 Br bromine	36 Kr krypton
5	37 Rb rubidium	38 Sr strontium	39 Y yttrium	40 Zr zirconium	41 Nb niobium	42 Mo molybdenum	43 Tc technetium	44 Ru ruthernium	45 Rh rhodium	46 Pd palladium	47 Ag silver	48 Cd cadmium	49 In indium	50 Sn tin	51 Sb antimony	52 Tr tellurium	53 I iodine	54 Xe xenon
6	55 Cs caesium	56 Ba barium	lanthanoids	72 Hf hafnium	73 Ta tantalum	74 W tungsten	75 Re rhenium	76 Os osmium	77 Ir iridium	78 Pt platinum	79 Au gold	80 Hg mercury	81 Tl thallium	82 Pb lead	83 Bi bismuth	84 Po polonium	85 At astatine	86 Rn radon
7	87 Fr francium	88 Ra radium	actinoids	104 Rf rutherfordium	105 Db dubnium	106 Sg seaborgium	107 Bh bohrium	108 Hs hassium	109 Mt meitherium	110 Ds darmstadtium	111 Rg roentgenium	112 Cn copernicium	113 Nh nihonium	114 Fl flerovium	115 Mc moscovium	116 Lv livermorium	117 Ts tennessine	118 Og oganesson
				57 La lanthanum	58 Ce cerium	59 Pr praseodymium	60 Nd neodymium	61 Pm promethium	62 Sm samarium	63 Eu europium	64 Gd gadolinium	65 Tb terbium	66 Dy dysprosium	67 Ho holmium	68 Er erbium	69 Tm thulium	70 Yb ytterbium	71 Lu lutetium
				89 Ac actinium	90 Th thorium	91 Pa protactinium	92 U uranium	93 Np neptunium	94 Pu plutonium	95 Am americium	96 Cm curium	97 Bk berkelium	98 Cf californium	99 Es einsteinium	100 Fm fermium	101 Md mendelevium	102 No nobelium	103 Lr lawrencium